Development and Implementation of Ship BWMS

Ning Chen · Zongming Yang · Wenbin Luo

Development and Implementation of Ship BWMS

SSTP

Springer

Ning Chen
Jiangsu University of Science
and Technology
Zhenjiang, Jiangsu, China

Zongming Yang
Jiangsu University of Science
and Technology
Zhenjiang, Jiangsu, China

Wenbin Luo
St. Mary's University
San Antonio, TX, USA

ISBN 978-981-33-6864-4 ISBN 978-981-33-6865-1 (eBook)
https://doi.org/10.1007/978-981-33-6865-1

Jointly published with Shanghai Scientific and Technical Publishers, China
The print edition is not for sale in China Mainland. Customers from China Mainland please order the print book from: Shanghai Scientific and Technical Publishers.

This Springer imprint is published by the registered company Springer Nature Singapore Pte Ltd.
The registered company address is: 152 Beach Road, #21-01/04 Gateway East, Singapore 189721, Singapore

Preface

The 2004 International Convention for the Control and Management of Ballast Water and Sediment of Ships (the "Ballast Water Convention") reached the conditions for entry into force on September 8, 2016, and took effect on September 8, 2017. The entry into force of the Convention will directly promote the market for ballast water management systems, bringing huge business opportunities to production enterprises.

Starting from the composition and working principle of the ship's ballast water management system, this book focuses on the characteristics and working principles of various types of popular ballast water management systems currently in the market, as well as the working principle, structural characteristics, calculation specifications, and calculation methods of simulation of various kinds of key components such as filtration, cleaning and sterilization in the system. The design examples of key components are given. From the perspective of system implementation, this book introduces the implementation method of ballast water management systems in new ship designs and old ship modifications. Starting from the operation management, the working principle of various detection sensors in the system and the control management method of the hardware man-machine interface in the control system are introduced. Finally, from the perspective of system maintenance, the maintenance and management methods of ballast water management systems are introduced, and the maintenance methods of various key components that make up the system are highlighted.

This book is suitable for (1) the personnel in charge of system development and design of ballast water management systems, and (2) the engineers and technicians in the area of system design for shipyards. It is also applicable to the crew members for equipment maintenance and management. In addition, this book can be used as a textbook for graduate students in marine engineering majors to further their professional studies.

Zhenjiang, China — Ning Chen
Zhenjiang, China — Zongming Yang
San Antonio, USA — Wenbin Luo

Contents

Chapter 1
Introduction

The 2004 International Convention on the Control and management of ballast water and sediment of ships (the "Ballast Water Convention"), which has been in discussion for 12 years, reached the conditions for entry into force on September 8, 2016 and took effect on September 8, 2017. For this matter, the China Classification Society (CCS) released the "2016 Technical bulletin No. 225" the latest technical bulletin. This technical bulletin interprets the Ballast Water Convention, summarizes the latest developments in IMO and the focus on the industry, and provides the latest documents issued by the United States in response to the USCG Ballast Water Act, covering all aspects of new shipbuilding design/construction, existing ship installation equipment/modification, and how to respond to, drawings, Inspection and certification, etc., with tips and suggestions. It provides the latest information and decision-making recommendations for shipowners, shipyard, design units, product manufacturers and other relevant customers.

The entry into force of the Convention directly urges the market for ballast water treatment systems, and large orders are emerging. For example, Qingdao Shuangrui Marine Environmental Engineering Co., Ltd. and Bohai Shipbuilding Heavy Industry Co., Ltd signed 13 210,000-load tons of bulk carrier ballast water treatment system contract, with the total contract price of nearly 70 million yuan (Chinese RMB); Wartsila received an order for 44 sets of ballast water treatment systems for 22 bulk carriers built by Yangzhou Guoyu Shipbuilding Co., Ltd., which is the largest single order in the ballast water treatment system market to date; China Ship Jiujiang Technology Research and Development Center signed a supporting contract for container shipsin August 2016 with a large shipping company, winning the first set of ballast water treatment system orders, and then signed a contract of providing the ballast water treatment systems for two 1700 TEU container ships owned by BussShipping, with the China Ship Huangpu WenChong Shipping Co., Ltd. Wuxi Blue Sky Electronics Co., Ltd. has also received the ballast water treatment system supply orders from Bohai Shipbuilding Heavy Industry Co., Ltd and Jiangsu Yangzijiang Group Corporation and other shipyards. The approaching date of entry into

N. Chen et al., *Development and Implementation of Ship BWMS*,
https://doi.org/10.1007/978-981-33-6865-1_1

force of the Ballast Water Management Convention, as well as this unprecedented wave of orders, heralds a real harvest season for the ballast Water treatment system market.

The enterprises, who entered into the ballast water treatment system market earlier, were busy in undertaking orders. Other enterprises, which were in an effort to pass certification and achieve industrialization and batch production, are to speed up the pace. Although there is always a hot topic in the ballast Water treatment system market, its market demand has not been fully demonstrated due to the failure of the ballast Water Management Convention to enter into force for many years. The enterprises, who produce Ballast Water treatment systems, have been doing more research and development, production base construction, market development and other investment work over the years. Even companies that have already received orders have limited financial returns in comparison with their investment. Only at a moment when the Ballast Water Management Convention is about to enter into force will these companies begin to receive financial "compensation" after waiting for many years.

For the upcoming ballast water treatment system market war, many ballast water treatment system enterprises are in a complex mood. But they all believe that in the face of great challenges, they see an even greater opportunity for development at the same time. The heads of many domestic ballast water management system production enterprises have expressed similar views: "With the entry into force of the Ballast Water Management Convention, all new ships and existing ships sailing across the sea are required to be fitted with ballast water treatment systems.In particular, nearly 50,000 ships currently in operation will be fully installed within the next 5 years. So, the installation of ballast water treatment systems for about 10,000 vessels per year needs to be carried out. There will be a blowout in the market for ballast water treatment systems, which will be a huge opportunity for the companies involved."

The huge market opportunities are getting closer and closer, but there are still some market obstacles and problems facing the ballast water treatment system enterprises. The first is the shortage of funds that is common to shipowners. The price of a ballast water treatment system is between US$50,000 and US$3 million. The current ship market is sluggish, the financing environment is harsh, and the shipowners with generally poor operating conditions are trying to fully reduce the expenditure and installing the ballast water treatment system is an additional cost increase. As a result, the shipowner will inevitably lack enthusiasm. Although the ballast water management convention is mandatory, how to let the shipowners who are short of funds take out the their money is still an issue that makes the ballast water treatment system manufacturers scratch their heads.

Secondly, the port state detection methods and standards for ship ballast water have not been introduced. Therefore, even if the ballast water treatment system approved by the classification society has been obtained, there is still a situation in which it could not pass the detection of the port state. If the forthcoming port state monitoring guidelines stipulate the traceability principle, how to deal with this risk by ballast water treatment system companies will be a big case.

At present, most of the ballast water treatment system manufacturers are optimistic about this, and believe that if the operation is carried out according to the requirements of the ballast water treatment system, it will be avoided to the maximum extent that the ballast water treatment system cannot be passed through the port state. They said that if the ballast water treatment system manufacturers fully understand the sampling methods and testing methods of the port countries and optimize the products in a targeted manner, and the shipowners also strengthen the training of their crew in the system operation, then the probability of obtaining the classification of ballast water treatment systems approved by the classification society while not passing the detection of port states will be very small.

Once again, shipowners on US routes are required to ensure that the ship complies with USCG regulations. Although many ballast water treatment systems have been approved by the USGS for AMS, according to the USCG regulations, the IMO-approved ballast water treatment system can be used for 5 years after being approved by the USCG AMS. The USCG will evaluate the feasibility of higher ballast water discharge standards and the results will be announced by January 1, 2016. This variable will cause numerous uncertainties in the ballast water treatment system, and relevant production enterprises may invest more human and material resources in research and development and production to meet the relevant standards.

All of these are the common problems faced by the global ballast water treatment system manufacturers. For Chinese companies, there are issues such as weak brand influence and relatively backward development of global after-sales service bases. Some industry insiders said that the single largest order received by Wartsila also confirmed this to a certain extent. Traditional leading companies of ship auxiliary supplies in Europe or U.S. have higher visibility and stronger market appeal. Therefore, China's ballast water treatment system manufacturers still have a lot of work to do in terms of brand building and global after-sales support.

Faced with a series of opportunities and challenges, China's ballast water treatment system manufacturers have made a lot of efforts and a lot of preparation work. In the face of the market demand that is expected to grow rapidly, more than ten companies in China have actively invested in R&D, production and marketing, and have done a lot of work in capacity building. At present, Jiujiang technology R&D center of CSSC is building a production base with capacity of 1000 sets of ballast water treatment system in Jiujiang economic and technological development zone in two stages. The project is supported by the state. The first stage of the industrialization of the annual output is 500 sets of units, and the center bought relevant patent technology from Wartsila. Qingdao Shuangrui has already invested in the production base with the capacity increased from the original 100 units to 500 units. Qingdao Haide Marine science and technology Co., LTD. will reach the production capacity of 2000 units in two years. Wuxi blue sky electronic Co., LTD formed the mass production in 2013 with support of Technology Department of Jiangsu provincial government. For this capacity building boom, the industry insiders think Chinese production enterprises of ballast water treatment system have already invested a huge amount of money in research and marketing. If the production is delayed, it means to give away the market. So, it is necessary to expand the capacity to seize market opportunities. But the industry of ballast water treatment system must also be alert to overcapacity. Once overcapacity occurs, the price war will be inevitable, and this will drag the industry into a vicious circle and the consequence is catastrophic. The industry insiders think Chinese production enterprises of the ballast water treatment system should adopt the unified business model as a group in the global market competition that is beneficial to strengthening themselves and enhancing their international competitiveness.

In the face of the entry into force of the Ballast Water Management Convention, the discharge of ballast water from operating ships must accept the risks that may occur in the port country sampling test. China's ballast water treatment system enterprises have also done a lot of work and formulated many plans. For example, domestic related companies have proposed to provide ballast water treatment system lease operation plans for ship owners around the world. The shipowner can obtain a customized ballast water treatment system only by paying a performance bond equivalent to 10% of the product price. After that, the shipowner pays the rent according to the contract every year. This kind of solution not only partially solves the problem of shortage of shipowners' funds, but also avoids the worries of ship owners.

For the potential uncertainty of ballast water treatment system, it is important to strengthen the timely communication between shipowner and production enterprise of ballast water treatment system. Some enterprises have already paid attention to and preparing for this. Shanghai ship research environmental protection technology co., LTD., for example, are preparing to build global salon center of ship ballast water equipment that provides a good communication, discussion and technical support platform for relevant parties such as ship owners, equipment suppliers, research and development institutions and the competent authority to jointly study and overcome practical difficulties in this industry.

In general, Chinese industry of ballast water treatment system is prepared for risks and challenges, but it should also be more actively involved in the formulation of the rules of the game. "In the inspection guideline of port state, for example, Chinese enterprises are only accepted passively, while foreign manufacturers lead the formulation by cooperating with the classification society and shipowners." After having researched the distribution of algae and marine environment along the coast of China, Professor Wu Hui-xian from Shanghai Ocean University sets up the species gene pool of the Chinese marine environment which provides the basis for related departments to participate in the international conference andto discuss and formulate inspection guidelines of port states, to prevent the guidelines becoming another tool of monopolizing the market of ballast water treatment system by only a few developed countries. At the same time, the government agencies are positively coordinating and guiding Chinese enterprises to strengthen their studying on ballast water treatment technology, in order to improve the technology level, so that the technical performance of their ballast water treatment systems always meetsor even reaches higher than the requirements of ballast water treatment standard. The shipowners should unite with equipment suppliers, classification society, scientific research institution and so on to act as soon as possible, and to pay attention to the operation of ballast water treatment system on board, and to maintain, sample, control, and monitor during the operation stage. This benefits not only the shipowner to prepare for management, operation, maintenance of ballast water treatment system as early as possible, but also benefits enterprises to accumulate experience about ballast water treatment system, and to speed up the upgrading of the product.

In addition, relevant government departments should also strengthen the guidance of supporting research and development of core components of ballast water treatment systems in China. At present, although there is almost no gap between the technology of China's ballast water treatment system and the international advanced level, the matching of core components is still subject to people. For example, the ballast water treatment system using ultraviolet method, its UV lamp and rectifier and filters and other products are basically dependent on imports. Therefore, in the face of global market competition, China's ballast water treatment system enterprises should not only strive to upgrade the products, but also strive to achieve the localization of core components. Only by making great efforts in product development and continuously improving the stability, reliability and economy of our products, can our company build a brand through products and become a winner in the upcoming ballast water treatment system market war.

Chapter 2
Ship Ballast Water Management System

Ship ballast water is a powerful means to provide stability for ships and to adjust the draft and draught of ships in order to meet the good handling performance. The maximum ballast capacity of a ship's ballast water is about one-third of the ship's volume, which makes about 10 billions tons of ballast water transfer between the various seas around the world every year [1].

2.1 Harm of Ship's Ballast Water

The sea life in the seawater of each sea area is different. These sea creatures are also transferred globally with the charge and discharge of ballast water. According to statistics, more than 3,000 sea creatures are transferred with ballast water every day. The harm caused by the transfer of sea creatures is terrible. First of all, there are some bacteria and viruses in sea life. If they can survive and reproduce, they will spread all kinds of diseases, and will bring great harm to the lives of local residents. Secondly, sea creatures contain various algae. When they are brought to a new "living area", where their natural enemies do not exist, the biological chain will be out of touch. As a result, these algae will multiply so much that their overpopulation in the nearby waters can be manifested by water blooms or red tides. To date, about 500 species of organisms have been confirmed to have been transmitted by ballast water intrusion [2, 3]. After statistics, IMO lists 10 major invasive marine organisms, as shown in Table 2.1:

N. Chen et al., *Development and Implementation of Ship BWMS*,
https://doi.org/10.1007/978-981-33-6865-1_2

Table 2.1 10 kinds of main invasive species

Name	Country of origin	Afferent	Harm
Zebra Mussel	Eastern Europe	Great Lakes region	Mass breeding has destroyed the local ecological environment and food chain, and seriously blocked the sewage outlets of buildings and ships, resulting in an economic loss of US $1 billion
Round Goby	Caspian Sea and Black Sea	Great Lakes region	Strong adaptive and reproductive ability to prey on local economic horns and antler eggs
Cholera bacteria	There are different kinds in different places	South America, Gulf of Mexico	Cholera was endemic in two ports in Peru in 1991, affecting 1 million people and causingmore than 1000 people to diein South america
Toxic algae	Large numbers of different kinds are stored in different places	Ballast water discharge area	Large numbers of blooms cause hypoxia and red tides, affecting local fishery production
European green crab	Europe	South Africa, South America, southern Australia	Adaptable, tough predator. Local crabs are replaced, disrupting local coastal ecosystems
Epicaridea	Black Sea	The Baltic	Mass reproduction plays a major role in local organisms, affecting fishery production
North American jellyfish	East coast of the United States	The sea of Azov, the Black Sea, the Caspian Sea	It is multiplied, feeds on zooplankton, destroys the local biological system, and caused serious economic losses to the fishermen of the Black Sea and the Azov Sea in the 1990s
North Pacific starfish	Japan and Alaska waters	Australia Tasmania	The breeding speed is fast, mainly feeding on shellfish including economical shellfish scallops and oysters, which seriously affects local fishery production

(continued)

Table 2.1 (continued)

Name	Country of origin	Afferent	Harm
Eriocheir sinensis	Northern Asia	In Western Europe, the Baltic Sea, West Coast	Breeding, burrowing holes in the dam, affecting the safety of the dam, while preying on local fish, affecting fishery production
Plum blossom envy fish	Eurasia	Great Lakes region	Mass reproduction, rapid growth, and strong adaptability have seriously affected local fishery production
European green crab	Europe	South Africa, South America, southern Australia	A strong, adaptable predator. It replaced the local crabs and disrupted the local coastal ecosystem

2.2 International Legislative Process for Ship Ballast Water

Problems caused by ballast water carrying harmful organisms were introduced at the IMO conference in 1973, when the meeting adopted a resolution and pointed out the seriousness of the problem. In 1990, the Marine Environmental Protection Committee (MEPC) set up a task force on ballast water at its thirty-first meeting to guide the drafting of invasive species. The guide was adopted at its fiftieth meeting. At the IMO conference in 1993, a number of investigations on ballast water were discussed, and new resolutions were adopted on the basis of the 1991 guidelines, giving new resolutions greater legal validity. In the second half of the twentieth century, the International Maritime Organization continued to improve the guidelines for carrying microorganisms on ballast water, and its effectiveness increased. In March 2000, MEPC held its 44th session. The continued improvement of the draft guideline by the Ballast Water Working Group became the focus of the conference.

Finally, the International Maritime Organization (IMO) adopted the International Convention on the management and control of ballast water and sediment (BWM Convention) in London in 2004 [4]. The Convention requires each country to take effective measures to deal with ships' ballast water and its deposits, and to allow port States to inspect foreign ships for the standards specified in the Convention and to impose sanctions and punishment on non-compliance emissions.

The Convention contains two parts: the main text and the supplementary. The supplementary section gives the ballast water performance standards and ballast water management standards, as shown in Tables 2.2 and 2.3.

In addition, the supplementary section also covers the approval procedures for ballast water treatment systems, namely, G8 type certification and G9 basic certification, and its specific process is shown in Fig. 2.1. According to the diagram,

Table 2.2 Ballast water performance standards

Standard name	Specific requirement				
D-1standard	*D*-1The standard is the replacement standard: the replacement rate of ballast water shall be at least 95%. For ships replacing ballast water by injection, the amount of water three times the volume of ballast tanks shall be deemed to meet the standards described				
D-2 standard	Viable microorganisms Indicators for organisms (including, but not limited to, the following three)				
	≥ 50 μm	≥ 10 μm and < 50 μm	Vibrio cholerae	Escherichia coli	Intestinal coccus
	< 10个/m^3	< 10个/mL	< 0.01 cfu/mL	< 2.5 cfu/mL	< 1 cfu/mL

it is not difficult to observe the use of pure physical methods to treat systems that do not produce any chemicals or other active substances. Only G8 type approval is required. A system that uses chemical additives or other active substances is required to undergo simultaneous certification by two processes, G9 and G8. As a result, certification of systems using chemical additives or active substances is cumbersome and the certification process takes place at the headquarters of the international maritime organization. In comparison, the former certification is relatively convenient, and the process can only be completed within the country. In short, from the point of view of system certification, the use of pure physical methods to deal with ballast water system certification cycle is short, saving development costs.

2.3 US Coast Guard Ballast Water Management Act USCG

The US Coast Guard issued the USCG for ballast water management regulations on March 23, 2012, and required to enter into force on June 21, 2012. This regulation is not subject to the entry into force of the IMO-related conventions, and its jurisdiction applies to ballast water discharge for all commercial vessels in the US waters. Commercial ships prior to U.S. waters must be installed with USCG approved ballast water management system, which is certified by the United States public water supply system (U.S. Public Water System) standards and requirements. The commercial vessels without the ballast water management system must be 200 nautical miles away from the nearest land, not discharging ballast water in the US waters, or discharging to the shoreline containment system or other ships. The first phase of the implementation of the USCG ballast water discharge standard is the same as the IMO D-2 standard (in New York). The second phase of the implementation of the standard has not been achieved due to current technical level, economy, environmental protection and other factors, so it will not be implemented until the implementation of specific programs in California and other states in the United

Table 2.3 Ballast water management standards

Year of construction	Ballast tank volumem^3	2008	2009	2010	2011	2012	2013	2014	2015	2016	2017
(2009, before)	[1500,5000]	$D-1/D-2$						$D-2$			
	< 1500, > 5000	$D-1/D-2$								$D-2$	
(2009, after)	< 5000	–	$D-2$								
(2009,2012)	≥ 5000	$D-1/D$-2								D-2	
(2012, after)	≥ 5000	–					$D-2$/Higher standards				

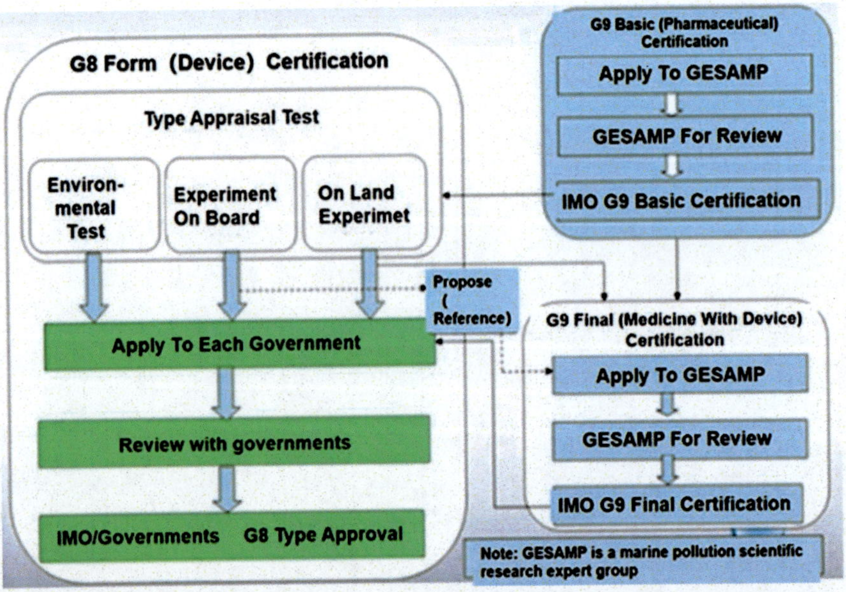

Fig. 2.1 Recognized flow chart of ballast water treatment system

States. The comparison of the IMO D-2 stage emission standards and USCG-stage 1, California standards and stage-2 is shown in Table 2.4.

The application of the USCG standard is shown in Table 2.5.

The US Coast Guard stipulates that commercial vessels, sailing in the US waters with a ballast water management system approved by other countries, must first apply for USMS AMS approval that will be available for 5 years from the applicable period. In the meantime, it is necessary to apply for the official type approval of the USCG. Usually, this process takes about 3 years. For ballast water management

Table 2.4 Comparison of IMO D−2 stage and USCG emission standards

Active object	IMO D−2	USCG-Stage 1	California standard	USCG-Stage 2
≥ 50 μm	< 10 per 1 ton	< 10 per 1 ton	Inactive organism	< 1 per 1 ton
≥ 10 μm < 50 μm	< 10 per 1 ml	< 10 per 1 ml	< 0.01 per 1 ml	< 1 per 1 ml
Vibrio cholerae	< 1 cfu per 100 ml	< 1 cfu per 100 ml	< 1 cfu per 100 ml	< 1 cfu per 100 ml
Escherichia coli	< 250 cfu 100 ml	< 250 cfu 100 ml	< 126 cfu 100 ml	< 126 cfu 100 ml
Intestinal coccus	< 100 cfu 100 ml	< 100 cfu 100 ml	< 33 cfu 100 ml	< 33 cfu 100 ml

Table 2.5 USCG standard application

Ship type	Ballast tank capacity	Construction time	Applicable time
New shipbuilding	Whole	2013.12.01 after	When the new ship is delivered
Shipping ship	1500 m^3 below and 5000 m^3 above	2013.12.01 before	2016.01.01 after the first docking
	1500–5000 m^3		2014.01.01 after the first docking

systems that have not yet been approved by other countries, if they are officially approved by the USCG, they can be installed on board without being restricted by the IMO D-2 standard.

References

1. IMO. 2000. GloBallast programme. London: Ballast Water News (1): 2–3.
2. Lewis, Patrick N., Chad L. Hewitt, Martin Riddle and Andrew McMinn. 2003. Marine introductions in the Southern Ocean: an unrecognized hazard to biodiversity. *Marine Pollution Bulletin* (46): 213—223.
3. Xue-feng, Wang. 2009. Marine ballast water pollution to the sea and its treatment method.*China Water Transport*9(1): 3–4.
4. IMO. 2004. International Convention for the Control and Management of Ships' Ballast Water and Sediments.

Chapter 3
The Working Principle of Ballast Water Management System

3.1 Ballast Water Treatment Methods for Ships

There are four methods for ship ballast water treatment: mechanical treatment, physical treatment, chemical treatment, and other treatment.

1. **Mechanical treatment method** [1]

    (1) **Filtration method**: When the ballast water is injected or discharged, the microorganisms and pathogens therein are removed after being treated by the filter. This method is very convenient and effective, but some viruses or bacteria are very small individuals, such as some viruses are only 0.02 μm in size. This brings a lot of trouble to the filtering method. If the filtration accuracy of the filter is increased, a high pressure is required at the time of filtration, and the filter is clogged, so that the cleaning frequency of the filter will also increase. Conversely, when the filtration accuracy of the filter is too low, the method will lose its meaning to the treatment of the ballast water. In summary, this method has problems in the feasibility of treating ballast water.

    (2) **Hydrocyclone separation method**: The main principle is to use the difference in specific gravity between the substances and achieve it by centrifugation. This method removes most of the microorganisms and turbidity from seawater. It has the advantages of simple operation and keeping the ballast tank clean. However, this method cannot be completed when dealing with microorganisms similar to the specific gravity of seawater [2].

    (3) **Dilution method**: When the ballast water needs to be discharged off-site, it is necessary to press the seawater farther away from the sea into the ballast tank, and at the same time discharge the water in the tank to the deep sea area. The continuous injection and discharge make the seawater in the ballast tank be diluted and it can be discharged at the port when the standard is reached. Obviously, the amount of seawater required by this

N. Chen et al., *Development and Implementation of Ship BWMS*,
https://doi.org/10.1007/978-981-33-6865-1_3

method is very large, and the volume of seawater displaced can be several times the volume of the ballast tank before the standard can be met.

(4) **High-speed flow method**: The treatment of ballast water is accomplished by mechanical damage to large microbial particles using a high speed, high pressure water pump. The disadvantage of this method is that the treatment is not thorough and the ballast water performance standard is not achieved; the cost of the equipment is greatly increased.

2. **Physical methods**

(1) **Heating method**: a method of sterilization by heating water in a ballast tank. The route of heating ballast water is usually by means of steam or the waste heat of a ship's engine. This process is very competitive, does not produce any pollution, and does not require too much extra processing equipment. However, due to the large amount of ballast water, this method will take a long time to deal with. Also, excessive temperature will cause some thermal strain, resulting in dangerous navigation.

(2) **Discharge**: this method can inactivate most of the microorganisms in ballast water. The disadvantages are as follows. The power consumption is too large. The spectrum of inactivation is not high. And there is a safety problem.

(3) **UV**: ultraviolet disinfection technology is to change or destroy the DNA structure of microorganisms through ultraviolet radiation, thereby killing bacteria or destroying their reproductive ability, and achieving the purpose of sterilization. According to the test, only C band of ultraviolet light has a real bactericidal ability, because UVC is easily absorbed by microbial DNA, which is the strongest bactericidal power at 254 nm wavelength. UVC sterilization is a physical method of disinfection, simple and convenient, high sterilization spectrum, no secondary pollution, easy to achieve automation, etc. With the introduction of various new UVC lamps, the application range of UVC sterilization has also expanded [3]. However, UVC disinfection can only kill microorganisms in the outer layer, and cannot inactivate microorganisms attached to solid particles or hidden inside spores. In the actual application process, due to poor water quality, high hardness, large bacterial content and other factors, it is easy to scale on the quartz tube wall, resulting in lower ultraviolet radiation energy, weakening penetration, and seriously affecting the sterilization effect. In addition, when the treated water leaves the reactor, some microorganisms killed by ultraviolet rays repair the damaged DNA molecules under the photoreactivation mechanism to regenerate the bacteria.

(4) **Ultrasonic**: Ultrasound mainly uses cavitation to sterilize. The periodic high-frequency vibration of the ultrasonic wave causes the micro-bubbles in the water to rapidly expand and then break, causing high temperature and high-pressure cavitation generated during the crushing and bursting process. Cavitation results in a series of chain reactions that can kill

harmful microorganisms in ballast water. However, this method also has many disadvantages, such as large energy consumption and limited water treatment.

(5) **Magnetization method**: this method has a strong lethal effect on calcareous invertebrate microorganisms, and has no obvious killing ability to other microorganisms. There are defects such as high cost and bactericidal limitations.

3. **Chemical methods**

(1) **Chlorine or chloride**: chlorine or chloride that uses its superior oxidizing properties to kill microorganisms in ballast water. When the solution concentration of chloride or chloride reaches a certain value, all microorganisms in ballast water can be killed. For example, chlorine or chloride solutions with concentrations greater than 10 mg/l can kill the most vigorous small crescent-shaped algae. This method often produces secondary pollution and causes corrosion to ballast tanks.

(2) **Hydrogen peroxide**: its bactericidal principle is the same as chlorine, and it is sterilized by strong oxidizing. Hydrogen peroxide (H_2O_2) in water can be hydrolyzed by the hydroxyl (OH^-), and hydroxyl has strong sterilization ability. Similarly, when the concentration of hydrogen peroxide reaches a certain value, all microorganisms in ballast water can be killed and even spores in the dormant period can be killed. However, the corrosion and cost of ballast tanks are bound to have limitations.

(3) **Metal ion method**: This method uses electrolysis to electrolyze copper and silver ions to achieve sterilization. This method is very successful in freshwater treatment. However, the disinfecting efficacy of such metal ions is also affected by the object to be treated and environmental conditions, such as the pH of the water to be treated or the resistance of the microorganism to the heavy metal ions. Although this water treatment method is much less toxic than chlorine or chloride solutions, the cost and power consumption of the method are very high.

(4) **Ozone treatment**: According to the molecular formula O_3 of ozone, it must have superior oxidizing power. According to the test, the redox potential can reach 2.07 V, and such a high voltage is enough to inactivate all microorganisms in the ballast water. Moreover, after the action of ozone and microorganisms, it will not cause secondary pollution.It is reported in the literature that ozone with a mass fraction of $4 * 10^{-6}$ can kill single-celled organisms and some resistant invertebrates. The cystscan be inactivated by ozone of mass fraction of $10 * 10^{-6}$ [4]. However, the long-term use of this method will corrode ballast tanks.

(5) **Hydroxy sterilization**: Hydroxy sterilization also utilizes its superior oxidizing properties. As long as the concentration reaches a certain value, all kinds of bacteria can be killed indiscriminately. At the end of the sterilization process, the hydroxyl groups combine with hydrogen ions to form

water without any secondary pollution. Since the hydroxyl technology is realized by electrolyzed water, the electric power consumption is high.

4. **Other treatment methods**
   Other non-mainstream treatment methods include: high seas replacement, electronic pulse and plasma pulse technology, deoxidation treatment, flocculation treatment, salinity treatment and organic biocides.In order to treat ballast water more effectively, people often combine these four methods to achieve the most efficient treatment of ballast water.

3.2 Sea Creature Preventing System for Ballast Water in China and Abroad

At present, the main ballast water biological management and development programs in China and abroad are as follows.

1. **Foreign management system**

   (1) **German Clean ballast management system**
   The system adopts the method of filtration and electrolysis disinfection, and does not add any active substances during the process. As a result, it only needs to meet the G8 certification process. The specific processing flow is shown in Fig. 3.1. After the ballast water is filtered, the larger microbial individuals are trapped, and the remaining smaller microorganisms come to the electrolysis disinfection module along with the ballast water, and the remaining microorganisms are deeply treated by the hydroxyl groups generated by the high-pressure electrolysis.

   (2) **The Hyde Guardian management system of the United States**
   The system employs filtering and ultraviolet radiation. The specific processing process is shown in Fig. 3.2. The ballast water is first filtered through a disc filter, and microorganisms larger than 50 μm are filtered off, and then irradiated with ultraviolet rays to kill the remaining microorganisms.

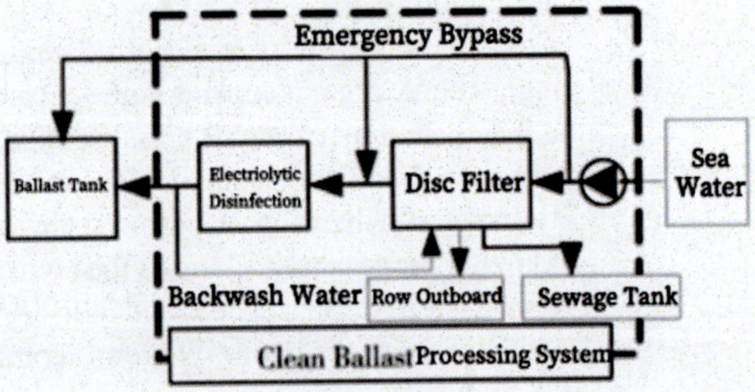

Fig. 3.1 Clean Ballast processing system [5]

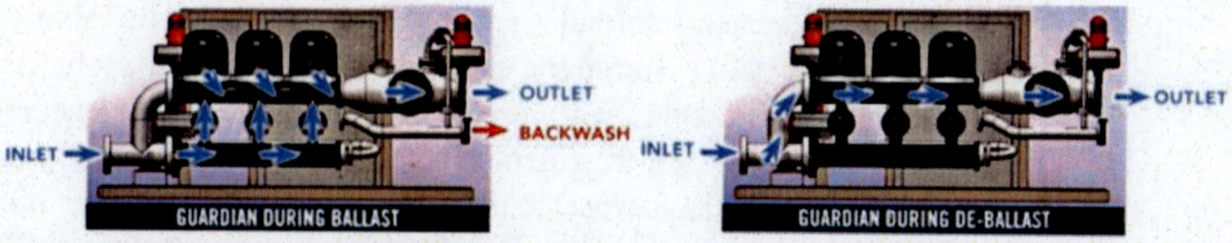

Fig. 3.2 Hyde Guardian processing system[6,7]

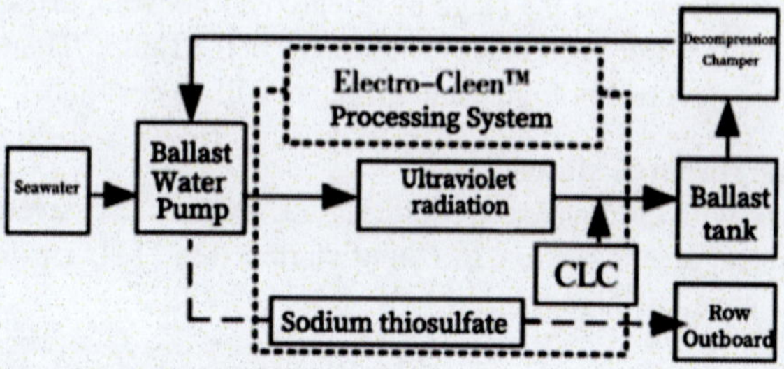

Fig. 3.3 Electro-Cleen™ processing system

(3) **Electro-Clean™ management system of South Korea**
The processing system includes a plurality of electrolytic modules. By electrolysis, seawater produces Cl^- and ClO^- and a small amount of OH [8], which kills microbes with the strong oxidation of these three ions. The specific flowof its processing is shown in Fig. 3.3.

(4) **PurBallast management system in Sweden**
The system is treated by filtration plus advanced oxidation [9, 10]. In this system, there are three processing modules: filter, AOT module, and CIP module. The system is special: ultraviolet radiation in the AOT module is mainly used to irradiate sea water to produce hydroxyl groups, and hydroxyl is the dominant factor in sterilization. The CIP module is mainly used to clean the AOT module. The processing flow is shown in Fig. 3.4.

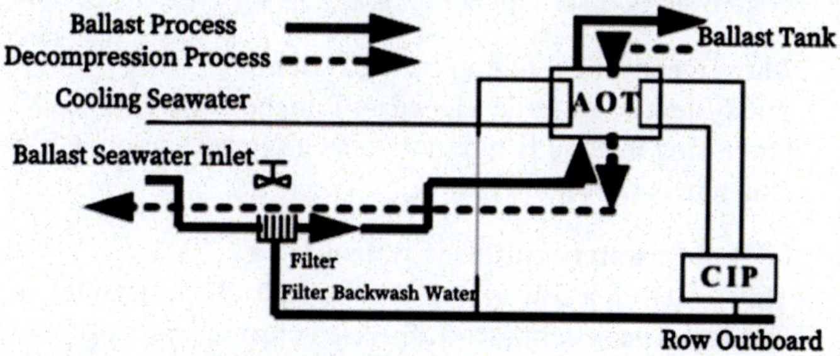

Fig. 3.4 PurBallast processing system

Other processing systems abroad are summarized as shown in Table 3.1. According to the above summary, most of the processing systems in foreign countries are combined with two or more processing methods [11]. This is because each treatment method has defects, so the single treatment technology cannot meet the requirements for ballast water treatment. From the authentication situation of foreign processing system, 50% of the final certification and initial certification are obtained, and the final certification is only 31% of the total. This fully illustrates that the technology is still in the stage of research and development. More and more manufacturers will enter the industry in the future. There is no doubt that more new technologies and products will be certified.

2. **Domestic research status.**
Compared with the developed countries, the research and development of ballast water management system in China started late [11]. Up to now, about 10 products have been recognized.

(1) **Qingdao ShuangRui BalClor™ management system**
The processing system was developed jointly by COSCO Group and institute of the CSIC725. As Shuang Rui group takes on the disinfection project of nuclear power plant for the year, its electrolytic chlorine technology is very mature. Therefore, the treatment method of BalClorTM management system is to make use of filtration + electrolysis to make chlorine + neutralization. The broadspectrum of chlorides is very high, but when the concentration is too large, the performance index of the ballast water exceeds the standard, so it is necessary to use sodium thiosulfate to neutralize it. The specific processing flow is shown in Fig. 3.5.

(2) **Wuxi blue sky BSKY™ management system**
The system is cyclonic separation + ultrasonic + ultraviolet. The workflow is shown in Fig. 3.6.

(3) **Qingdao Hadeway Ocean Guard management system**
The processing method of the system is filtering and ultrasonic. Using ultrasonic cavitation to produce hydroxyl to kill microorganisms, its system structure is shown in Fig. 3.7.

(4) **Blue Ocean shield management system**
The system is jointly developed by Tsinghua University and COSCO. The processing method is filtration + swirling separation + ultraviolet. The structure is shown in Fig. 3.8.

(5) **Ni Ballast water treatment system**
This system is a new system developed by Jiangsu Antarctic Machinery Co., Ltd. The system uses a 50μmself cleaning filter to remove the coarse

Table 3.1 Summary table about BWMS and certification at abroad

Name	Research and development countries	Processing method	Certification status
Clean Ballast	Germany	Filter + sea water electrolysis	Final certification
SiCURETM	Germany	Filter + *NaClO*	Preliminary certification
SEDNA	Germany	Swirl separation + filter + Biological killing	Final certification
Ocean Protection	Germany	Filter + *UV*	
Electro-CleenTM	The Republic of Korea	Sea water electrolysis	Final certification
NK-O_3 Blue Ballast	The Republic of Korea	O_3	Final certification
Blue Ocean Guardian	The Republic of Korea	Filter + *UV*	Preliminary certification
EcoballastTM	The Republic of Korea	Filter + *UV*	Final certification
Hiballast	The Republic of Korea	Filter + sea water electrolysis	Preliminary certification
En-ballast	The Republic of Korea	Filter + sea water electrolysis	Preliminary certification
GloEn-Patrol	The Republic of Korea	Filter + *UV*	Final certification
Special Pipe Hybrid	Japan	Filter + O_3	Preliminary certification
Clear Ballast	Japan	Flocculation + magnetic separation + filter	Final certification
JFE-BWMS	Japan	Filter + *NaClO* + cavity	Final certification
ATLAS BWTS	Denmark	Filter + sea water electrolysis	
DEMSI Ocean Guard	Denmark	Filter + *UV* + O_3	Preliminary certification
Greenship's Sedinox	Netherlands	Hydrocyclone + sea water electrolysis	Final certification
EcochlorTM	U.S.A	Filter + ClO_2	Preliminary certification
Ferrator	U.S.A	Ferrate	
BW treatment system	U.S.A	Deoxidation + cavity	
BWDTS	U.S.A	O_3 + Ultrasonic	

(continued)

Table 3.1 (continued)

Name	Research and development countries	Processing method	Certification status
Hyde Guardian	U.S.A	Filter + *UV*	
VOS	U.S.A	Deoxidation + cavity	
Maritime solution inc	U.S.A	Filter + *UV*	
Balpure	U.S.A	Sea water electrolysis	Final certification
Aquatic enhancement	U.S.A	Filter + *UV* + O_3	
ModelEL1-3B	U.S.A	Sea water electrolysis	
MARENCO	U.S.A	Filter + *UV*	
Brilliant marine LLC	U.S.A	Electronic pulse	
SCX2000, Mark3	U.S.A	O_3	
Sedinox	Finland	Filter + UV	Final certification
Crystal ballast	Finland	UV	
Ocean Saver BWMS	Norway	Filter + Electrodialysis + Cavitation	Final certification
Optimarin Ballast	Norway	Membrane technology + UV	
SeaSafe-3	Australia	Heating	
Unitor BWTS	South Africa	Cavity + O_3 + sea water electrolysis + filter	Final certification
MEXEL	France	Biological killing agent	
Ballastmar	Spain	Filter + sea water electrolysis	
Pure Ballast	Sweden	Filter + Ultraviolet/TiO_2	Final certification
Coldharbour Marine	Britain	Deoxidation	
Atg UV Technology	Britain	Swirl separation + filter	
SEDINOX	Britain	Swirl separation + sea water electrolysis	

particles first, and then removes all microbes and even bacteria by micro-filtration membrane and is filled with inert gas-nitrogen gas. The structure is shown in Fig. 3.9.

(6) **Other domestic management systems**
The ballast water management system of Dalian Maritime University, the ballast water management system of Jiangsu University of Science

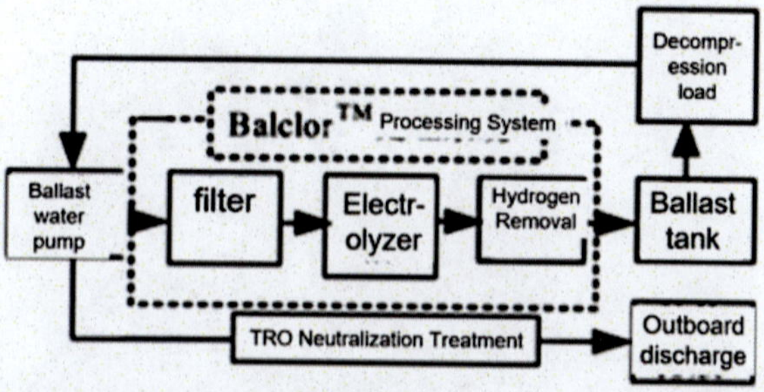

Fig. 3.5 BalClor™ processing system

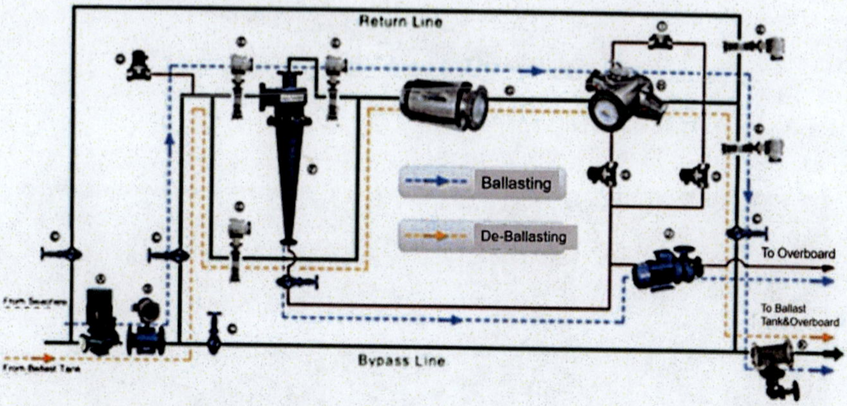

Fig. 3.6 BSKY™ processing system

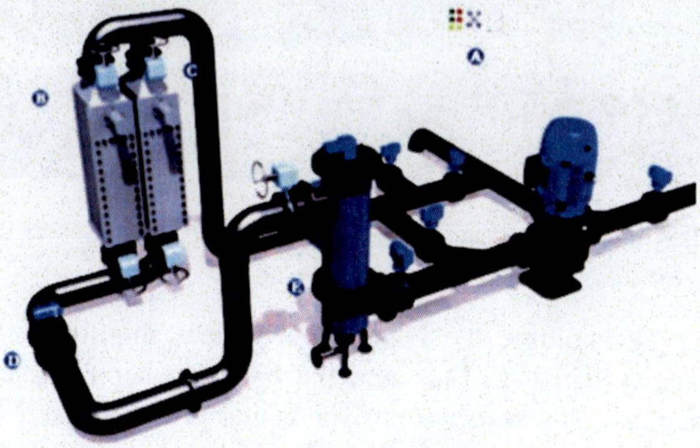

Fig. 3.7 Ocean Guard processing system

Fig. 3.8 Blue Ocean shield processing system

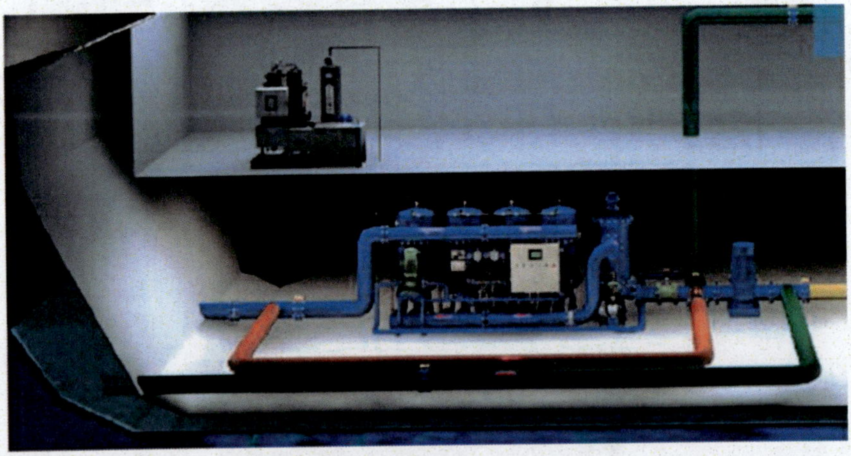

Fig. 3.9 Ni ballast water processing system

and Technology, the Nantong sea view ship ballast water management system Co., Ltd., and the ballast water management system of Shanghai ship research environmental protection and Technology Co., Ltd. Its processing principle is also the combination of several processing methods. The former is bactericidal by filtration + hydroxyl radical, while the latter is filtering + ultraviolet (+ ultrasound).

Thus, in order to meet the requirements of IMO and USCG, multiple filter and bactericidal techniques are used in the integrated development of the systemto achieve the best effect of preventing the propagation of marine life with ballast water. The filter equipment of the system consists of the backwash filter, the swirling filter, the cleaning equipment, the ultrasonic cavitation purifier, the ultrasonic cleaner, the UVC germicidal device, the electrocatalytic sterilizer, the electrolysis or the drug sterilizer, and the inert gas sterilizer.

References

1. Yong-xin, Song and Dang Kun et al. 2002. The measures of ships' ballast water. *World Shipping* 26(2): 42–43.
2. Armstrong G, Brooeks E, and Lankester R et al. 1997. Ballast system design for through exchange of ballast water. *Transactions-Institute of Marine Engineers* 109(3): 257–269.
3. Tian-tao, Zhu and Gong Xiao-yong et al. 2008. UV sterilization technology. *Journal of Chifeng University* (Natural Science Edition), 24(2): 23–24.
4. Guo-fang, Hu. 2008. Ecological impact and treatment technology of ship ballast water. *China Water Transport* (9): 48–49.
5. Kai, Zhang and Fang Zhen-dong et al. 2012. The research progress in ballast water treatment system. *Environmental Science Survey* 31(1): 49–52.
6. Wright, David A., Robert W. Gensemer and Carys L. Mitchelmore et al. 2010. Shipboard trials of an ozone-based ballast water treatment system. *Marine Pollution Bulletin* 60(9): 1571–1583.
7. Gollasch, Stephan, Matej David, Matthias Voigt, Egil Dragsund, Chad Hewitt and Yasuwo Fukuyo. 2007. Critical review of the IMO international convention on the management of ships'ballast water and sediments. *Harmful Algae* 6(4): 585–600.
8. David, Matej, Marko Perkovič, Valter Suban and Stephan Gollasch. 2012. A generic ballast water discharge assessment model as a decision supporting tool in ballast water management.*Decision Support Systems* 53(1): 175–185.
9. Tsolaki, Efi, Paraskevi Pitta and Evan Diamadopoulos. 2010. Electrochemical disinfection of simulated ballast water using Artemia salina as indicator. *Chemical Engineering Journal*156(2): 305–312.
10. Ballast Water Development (One) [EB/OL]. 1746. https://www.hyqb.sh.cn/pulish/portalo/tab140/infor.
11. Tamburri, Mario N., and Kerstin Wasson. 2002. Ballast water deoxygenation can prevent aquatic introductions while reducing ship corrosion. *Biological Conservation* (103): 331–341.

Chapter 4
The Key Components of the System and the Working Principle of the Unit

Check for updates

4.1 The Working Principle of the Filter Equipment

The function of the filtering equipment is mainly to filter the ballast water entering the ship from the sea chest, so that the sediments and algae spores, hyphae, pathogens, etc. carried by the water with a size of more than 50 μm are filtered out. The sterilizer can exert optimal bactericidal performance, and kills microscopic spores, hyphae, pathogens and the like of microalgae which are less than 50 μm contained in the filtered seawater. Commonly used mechanical filtration equipment includes backwash filters and swirl filters.

1. **The working principle of backwash filter**

   Backwash filtration is to enter raw water from the inlet of the filter, and then filter the clean water out of the filter outlet from the inside to outside. The impurity accumulates on the inner surface of the filter to form a filter cake to make the pressure difference inside and outside the filter. When the pressure difference reaches the preset value set by the differential pressure switch, the self cleaning process will be carried out [1].

   The backwash is done by a scanning sucking mechanism moving along the inner surface of the filter. Its structure is shown in Fig. 4.1. Sucking scanners, also known as suction devices, are hollow structures, and several suction pipes are evenly distributed with a certain distance along its axis. The cleaner is connected with a drain valve. When the drain valve is opened, the pressure difference formed between the internal water pressure of the filter and the external atmospheric pressure causes the suction pipe to generate strong suction force, and the water flows from the outside of the filter screen to the inner high-speed reverse flow to flush the impurities, and enters the suction device through the drain valve to discharge. The cleaner is driven by a bi-directional motor with a deceleration mechanism for a spiral motion at a fixed speed, so that a plurality of suction nozzles will be able to absorb the entire surface of the filter. During the whole anti cleaning process, the limiting device on the wire rod moves up

N. Chen et al., *Development and Implementation of Ship BWMS*,
https://doi.org/10.1007/978-981-33-6865-1_4

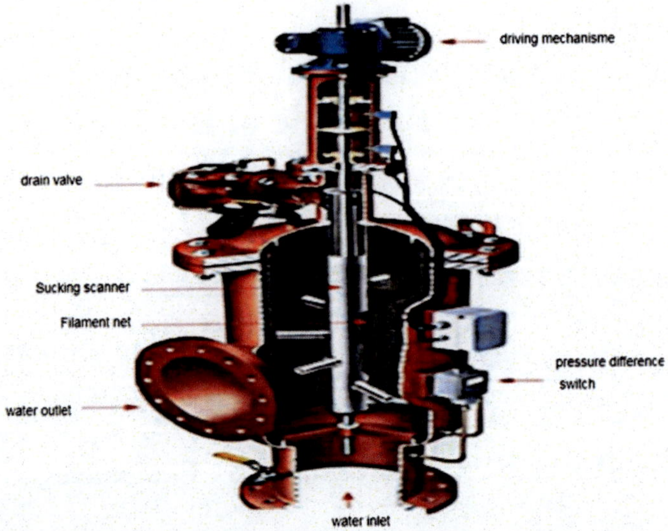

Fig. 4.1 Structure of self-cleaning filter

and down under the traction of the wire rod. When the upper and lower limit switches are encountered, the two-way motor runs in reverse direction until the internal and external pressure difference of the filter tends to be equal, and the motor stops driving. The system does not break off the flow during backwashing, and the whole process of the filter is controlled by an automatic controller [2]. Therefore, this filter is usually called a backwash filter.

2. **The working principle of the swirl filter**

The swirl filter is based on the principle of centrifugal sedimentation and density difference. When the ballast water enters a swirling flow filter at a certain speed along the tangential inlet, the hig speed rotating flow field is formed in the column section. The rotating flow field produces centrifugal action. The large density particles and aquatic organisms move down to the outer and axial direction along the radial direction, and the flow is from the bottom outlet flow. Under the action of anti vortex, the water with smaller density moves along the center axis and flows out of the overflow outlet to achieve the effective separation of solid–liquid two phases [3]. The structure of the hydrocyclone, as shown in Fig. 4.2, mainly includes the main diameter D of the cyclone, the inlet structure and the diameter D_i, the overflow diameter D_0, the bottom diameter D_u, the cone angle θ, the insertion depth of the overflow H0, and the length H_c of the column section [4].

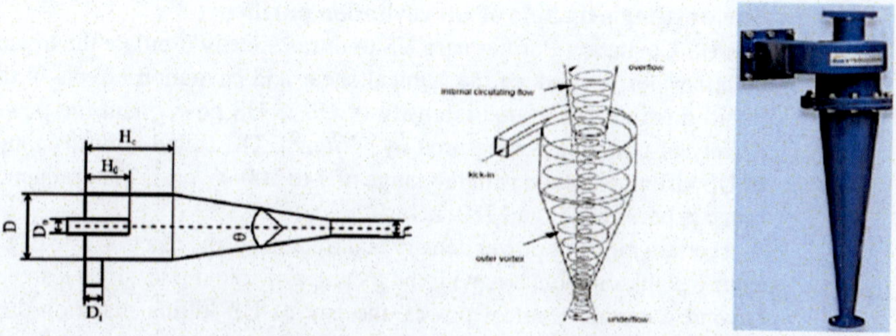

Fig. 4.2 Structural parameters of Hydrocyclone

4.2 Working Principle of Ultrasonic Cleaner

The ultrasonic (US) cleaner is a container with US participation and reaction under its action. It consists mainly of a US transducer, a US generator and a processing container [5]. The processor chooses a specific frequency to launch a sound wave for instant treatment of microbiological cells. When the US washer handles ballast water, the treated water enters the processor from the inlet and receives the US radiation treatment in the chamber. Finally, it is discharged from the outlet. In the process of processing, we mainly use the following functions of US:

1. **Mechanical effect**
   Mechanical effect is the most basic primary effect of US. No matter how much US intensity is, it will produce this effect. During the propagation process, US will cause the simple harmonic motion of the particle in the medium. In the process of the particle vibration, the acceleration is proportional to the square of the vibration frequency of the US, and can even reach the gravitational acceleration of tens of thousands of times, thus producing the mechanical effect [6] to the medium, and the instantaneous shock wave shear stress can reach 50 MPa, in this shear force. The shells of fungi or algal cells will be shredded or even killed directly by mechanical shearing.

2. **Cavitation effect**
   US cavitation is another unique physical process produced by the action of US in liquid medium. Under the action of US, small bubbles are produced in the liquid and vibrate under the continuous action of US. When the sound pressure reaches a certain value, these bubbles will expand and close quickly and produce a micro jet that is higher than 100 m/s in the process of closure. It finally collapses and releases the heat of 2000 –5000 °K. The cavitation bubble collapses and contracts, the temperature change rate reaches 109 K/s, and there is a strong shock wave in homogeneous liquid medium. The dynamic process of vibration, expansion, closure, collapse, and release of heat is called ultrasonic cavitation [6].

(1) **The working principle of US cavitation purifier**

The US Cavitation Purifier uses US to ultrasonically oscillate the treated ballast water to produce mechanical shear and cavitation effects in the liquid medium.The sound intensity of US is US power, and the power density of unit area is measured by (W/cm^2). The sound intensity range of US killing plankton is in the range of 1 to 100 W/cm^2. The frequency range is between 20 and 750 kHz.

In a certain range of power density and frequency, the reaction rate of US killing plankton increases with the increase of sound intensity. However, beyond a certain level of power, the rate of US killing plankton does not increase any more. The reason for this phenomenon is that under the action of large sound intensity, the cavitation bubble grows too much in the negative sound pressure phase, resulting in the compression of the positive sound pressure phase that is not enough to be collapsed.Under the action of large sound intensity, a large number of cavitation bubbles are activated, which produce strong scattering attenuation for US, and the ability to kill plankton is no longer increased.

The excessive frequency will make the cavitation bubble circulate too fast and affect the collapse of cavitation bubbles. Under the same sound intensity, the increase of the general frequency is beneficial to kill the plankton and algal spores in ballast water.

However, when a certain sound US enters the water body of the ballast water, the negative pressure phase causes the cavitation bubble in the water body, and the cavitation bubble is contracted under the positive pressure of the US. The surface rate of the cavitation bubble is faster than the sound velocity, and the process only lasts for a few microseconds, thus producing instantaneous high temperature and high pressure at this point, that is the so-called hot spot [7]. The water vapor entering the cavitation bubble has a splitting and chain reaction at high pressure and high temperature, producing OH, and the shock waves and jets produced by the cavitation bubble cause OH to enter the whole ballast water to accelerate the oxidation of plankton in the ballast water. The US cavitation purification reaction of plankton in ballast water includes the pyrolysis reaction and oxidation reaction, in which the hydrophobic spores cell wall breaks down under the impact of the cavitation collapse, and the cell membrane enters the cavitation bubble under the action of the jet, and the pyrolysis reaction is carried out. The cytoplasm is oxidized by the cavitation.

The main factors affecting the US gas purification reaction of plankton are: the structure of the reactor, the physical parameters of the US field, and the saturated gas in the ballast water and the solid particles of the plankton spore.

Ozone and electrolysis cannot be used separately in the world at at least 75 major ports, such as Rotterdam port in Holland (the third busiest port in the world), Antwerp port in Belgium (the sixth busiest port in the world), Hamburg Port in Germany (the nineteenth busiest port in the world), the port of the European Baltic Haiti, and the 24 main ports including Shanghai port along the Yangtze River in China, the port of Shenzhen, Ningbo port, the New Orleans port of the United States, the most bustling port area of the Mississippi River, and the Philadelphia port.

The US cavitation reaction, if designed inappropriately, produces a large number of OH, which is like the oxidation reaction between the ozone and the ballast water. Therefore, when designing the US cavitation purifier, the selection of the US transducer should be taken into full consideration. It is necessary to take advantage of the mechanical shear effect of US, and to consider the cavitation of US, to produce the necessary shock waves and jets. It shatters the spores of hard algal spores and prevents the production of large amounts of OH during this process. Usually, the frequency of the US transducer is too low, the cavitation is too strong, and the cavitation is easy to occur. It causes corrosion and damage to the processor itself and reduces the life of the processor. The frequency is too high, and the cavitation effect is too weak. The mechanical shear equivalent of the cavitation is weakened, and the processing effect of the processor is not good. In general, the transducer frequency 28 KHz can meet the above requirements. In addition, the size of the end face of the transducer selected by the US processor is also related to the processing efficiency, and the end face is large, which is more conducive to the improvement of the US radiation, so that the treatment effect is better [8]. Therefore, for the structure of the US processor, the transducer with larger radiation area should be chosen under the same design power. The structure of the US transducer is shown in Fig. 4.3 andthe structure of the US cavitation purifier is shown in Fig. 4.4a, b, c.

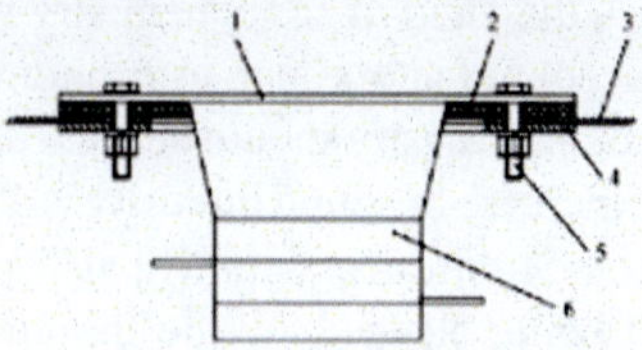

1. The vibration plate, 2.Damping, 3. The bottom wall of response,
4. Flange, 5. Bolt, 6.US transducer

Fig. 4.3 The structure of the flange connected US transducer

(a) Shape of the purifier, (b)Arrangement of the US vibrator, (c)Internal channel

Fig. 4.4 Structural decomposition of ultrasonic cavitation purifier diagram

(2) **Working principle of ultrasonic cleaner**

Unlike the ultrasonic cavitation purifier, the ultrasonic cleaner is not installed as an independent device in the ballast water treatment system, but uses its cleaning capability to be designed in parallel with a key device backwash filter or UVC sterilizer in the ballast water management system, combined to promote adhesion to backwash filtering. And the UV lamp of the UVC sterilizer is continuously cleaned to ensure that the quartz glass sheath of the UVC lamp with higher temperature is not fouled, thereby ensuring that the UVC is effectively irradiated onto the treated ballast water.

From the point of view of improving the efficiency of electric sound, it is necessary to match the sound intensity I with the over-current cross-sectional area of the treated ballast water in the design of the cleaner. According to the related design experience, the vertical depth of ultrasonic radiation should not be more than 400 mm when the detergent is treated with the processed ballast water, and the installation form of the ultrasonic transducer is on the vertical surface of the over current. At this moment, the power of the ultrasonic wave should be designedaccording to the area power density parameter (sound intensity I). Within a certain limit, increasing the sound intensity I will increase the yield of ultrasonic cavitation purification reaction, that is to say, produce excessive OH. Under normal circumstances, when the sound intensity I is greater than 3 W/cm^2, the yield of sonochemistry will not increase significantly. Taking into account factors such as economy and service life of cleaning equipment, sound intensity should be controlled within a certain range. For example, the increase of sound intensity will increase the power consumption of ultrasonic transducer, and will also have adverse effects on the life of the object being cleaned. While the ultrasonic transducer used with a small end face has a strong value in the ballast water treated in some areas, the other part weakens the sound intensity of the ultrasonic radiation, which will not improve the rate of ultrasonic cavitation purification, but it reduces the effect of ultrasonic cleaning on the whole device. Therefore, from the viewpoint of suppressing the ultrasonic cavitation purification reaction yield and reducing the power consumption of the ultrasonic transducer, the end face area of the transducer selected by the ultrasonic

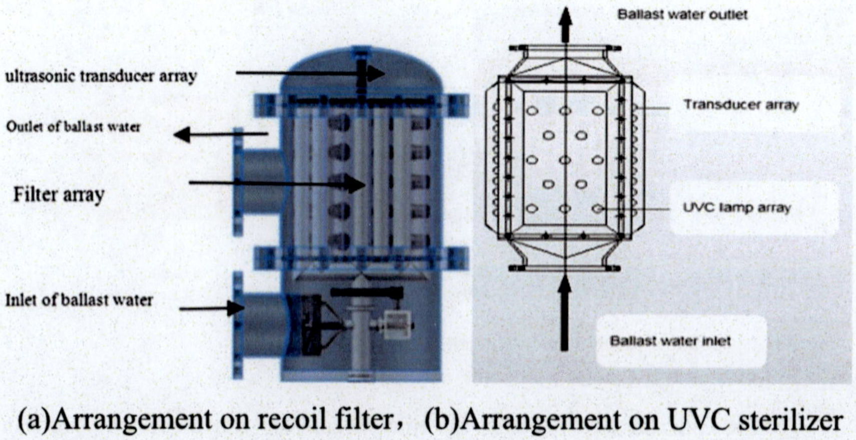

Fig. 4.5 The arrangement position of the ultrasonic cleaner

cleaner should be as large as possible, making the average sound intensity $I \leq 3$ W/cm^2. The ultrasonic transducer is arranged on the recoil filter and sterilizer as shown in Fig. 4.5a, b.

4.3 The Working Principle of Sterilizer

The UV sterilizer is the most important device for killing plankton in the integrated development of ballast water treatment system. Because it uses the bactericidal characteristics of the C band of ultraviolet ray to kill the plankton in the ballast water, this kind of bactericide is also called the UVC germicide.

1. **The working principle of UVC sterilizer**

   In 1801, German physicist Ritter discovered that a section of sunlight spectrum could make photographic negatives containing silver bromide photosensitive, and thus discovered the existence of ultraviolet radiation. Ultraviolet (UV) is the electromagnetic wave with wavelength between visible light (400 nm) and x-ray (100 nm). According to the different biological effects, ultraviolet radiation can be divided into four bands, UVA, UVB, UVC, and vacuum ultraviolet, as shown in Fig. 4.6.

   UVA (400–320 nm): can make human skin produce melanin, make skin darkening.

   UVB (320–275 nm): it can cause cancer and wrinkle the skin.

   UVC (280–200 nm): all is absorbed by the atmosphere, usually only by artificial light source. The UVC of its wavelength between 240 nm and 270 nm can directly destroy the DNA (deoxyribonucleic acid) and RNA (ribonucleic acid)

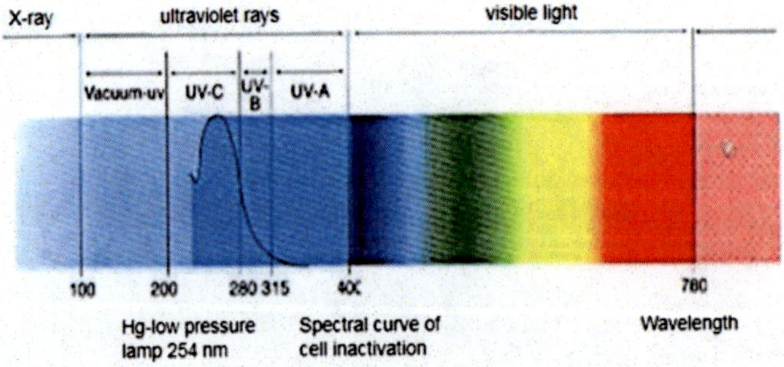

Fig. 4.6 Spectrum of theultravioletradiation

of the cell and virus, so that the microorganisms can die quickly and have bactericidal efficacy.

Vacuum ultraviolet (200−100 nm): can not enter the atmosphere, exist in space. The main way for the interaction of UVC with the matter is to make the electrons of the material transition from the ground state to the excited state, and UVC photons themselves are absorbed as energy by atoms or molecules of the matter [9]. It is generally believed that UVC mainly activates some photosensitizing molecules to induce free radicals and lipid peroxidation by direct radiation to damage cells of organisms, etc. The inactivation effect of UVC on marine organisms and microorganism is mainly achieved through the first two ways, that is, UVC direct radiation and photochemical reaction [10]. On the one hand, nucleic acid, the basic component of the body of life, has a strong absorption peak at the wavelength of ultraviolet radiation, that is, at the 253.7 nm band. When the nucleic acid absorbs ultraviolet radiation up to a certain dose, DNA will change, resulting in the synthesis of proteins and enzymes in the microorganism, and on the other side, free radicals induced by UVC cause photoionization, leading to cell death [11]. Moreover, the inactivation effect of UVC radiation has an "accumulation" effect.That is, when one radiation cannot kill the microorganism, it can be achieved by multiple radiations.

Irradiation by UVC destroys or changes the structure of microbial DNA, causing some bacteria to die directly, while others do not reproduce their offsprings to achieve sterilization. The principle of sterilization is shown in Fig. 4.7.

Fig. 4.7 Schematic of UVC germicidal

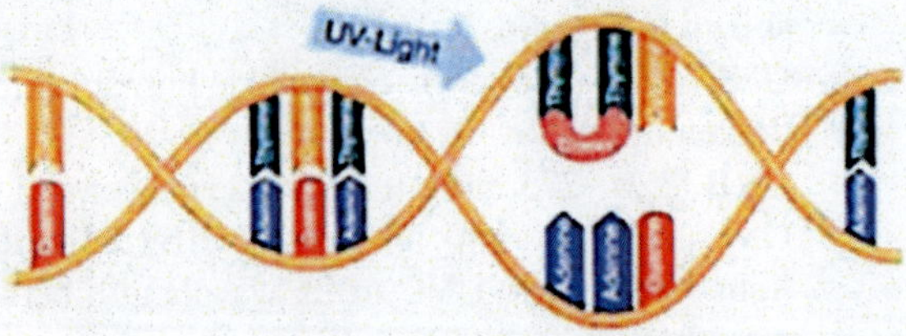

The mechanism of bactericidal action of UVC mainly includes the following points: ① It can make the nucleic acid mutation, and then prevent its reproduction, transcription, and other reproduction activities; ② It can directly damage the cell, inhibitting the activity of the cell; ③ It can produce free radical, thus killing the bacteria directly.

UVC sterilizer is a way to kill plankton in ballast water by converting electric energy into UVC light energy. According to the different ways of converting electric energy into UVC light energy, the device is divided into three types: low pressure UVC sterilizer, medium high voltage UVC sterilizer, and microwave UVC sterilizer.

Low pressure UVC sterilizer: Using low voltage amalgam ultraviolet lamp, the single root power of the lamp is about 300 W; the output rate of ultraviolet energy conversion is above 30%; the service life is 1.3 to 16 thousand hours, and the UVC lamp with long life and high conversion power. It is insensitive to temperature fluctuations and can be used at ambient temperatures of up to 90 °C. In addition, the unique coatingensures that they do not occur transmission loss of quartz glass, so in the whole life period of the lamp, it will have a continuous and stable bactericidal effect. The tube shape is shown in Fig. 4.8, and the output spectrogram is shown in Fig. 4.9.

When the low pressure amalgam UVC lamp is used as a bactericidal light source in the ballast water management system, because of its low power density, it needs to get the necessary UV dose to kill the plankton in the process of the flow of the ballast water from the inlet of the germicide into the outlet of the germicide, so that the bactericidal effect can be achieved. In the design of the UVC sterilizer, there are many sets of lamp tubes, which greatly increase the volume of the sterilizer.

The advantages of this kind of sterilizer are: the cost of the lamp tube and the driving power supply is low; the length of the lamp tube is long (can reach 1.5 m), so the flow rate of the ballast flow is reduced, the time of the dose is long by the UVC irradiation, and the life of the lamp tube is long. The drawback is

Fig. 4.8 Low pressure UVC lamp

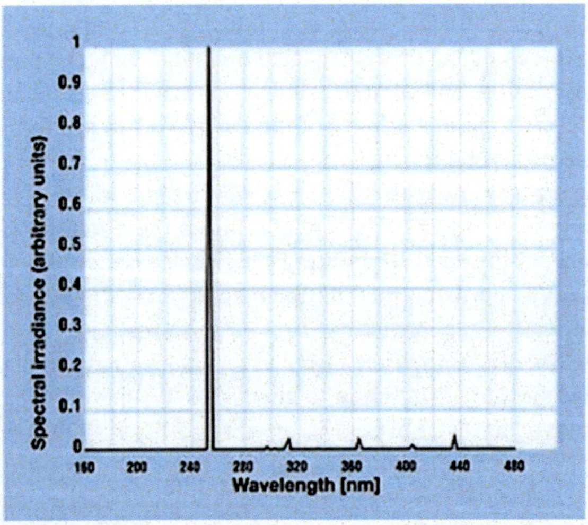

Fig. 4.9 Amalgam light spectra

that the required size of the germicide is larger. The structure of the low pressure and high output UVC apparatus is shown in Fig. 4.10.

Medium pressure high strength UVC sterilizer: using medium pressure high strength UVC lamp, the single tube power of the lamp is up to 2500 W, the UV conversion rate is 400 W, and the surface temperature of the lamp is between 650–900 °C when working normally. Compared with the low pressure high-strength amalgam UVC lamp, when used as a bactericidal light source in the ballast water management system, the ultraviolet energy output of the lamp is strong, the penetrating capacity of the water body is high, and even the turbid water with low penetration rate can be completely treated. Therefore, it is suitable for the ballast water treatment system of large flow. The shape of the tube

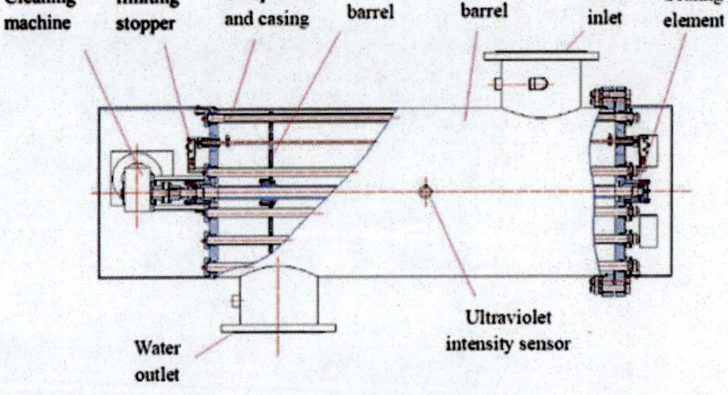

Fig. 4.10 Assembly drawing of UVC Low voltage high output UVC sterilizer

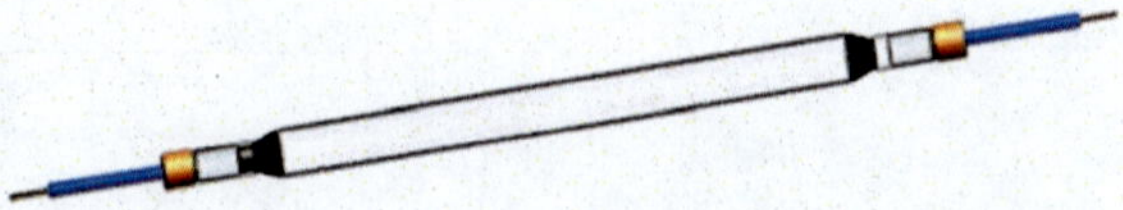

Fig. 4.11 Structure of DQ3030 medium pressure UVC lamp

Table 4.1 DQ3030-specific parameters of the medium pressure UVC lamp

Tube length	Arc length	Power	Tube voltage	Lamp current	Installation length	UVC output power
22 mm	300 mm	3000 W	415 V	7.5 A	420 mm	470 W

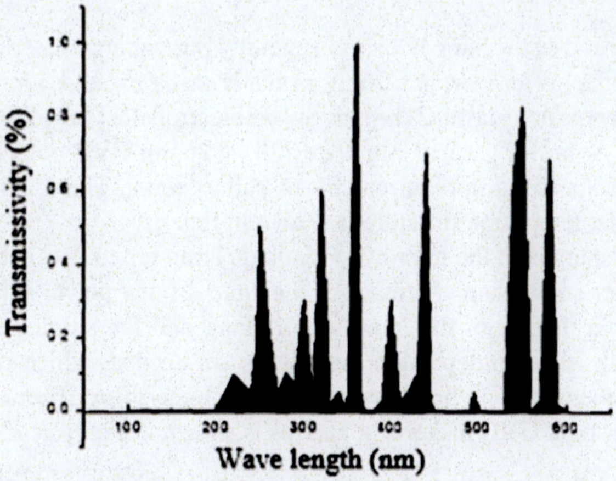

Fig. 4.12 Light spectrum of medium-pressure and high-strength UVC lamp

is shown in Fig. 4.11. The parameters are shown in Table 4.1, and the output spectrum is shown in Fig. 4.12.

The advantage of medium-pressure high-strength UVC bacteria device is that it has the strongest bactericidal ability and the size could be far shorter than the low-pressure lamp. It meets the requirements of the UVC apparatus with very large processing flow and the strict requirement of the equipment space. The drawback is: the electricity consumption is very large, the UVC conversion rate is less than 20%, the service life is lower than the low voltage lamp, and the matching cost of the driving circuit is high. The structure of medium-pressure high-strength UVC bacteriadevice is shown in Fig. 4.13.

Microwave UVC sterilizer: using the Microwave Discharge Electrodeless UVC Lamp (Microwave Discharge Electrodeless UVC Lamp, MDEUVCL), the microwave energy of 300 MHz to 300 GHz is converted into the internal energy of the gas molecule, which can be excited, ionized, and decomposed to produce the gas dischargeof the plasma [12, 13]. The microwave non-polar UVC lamp

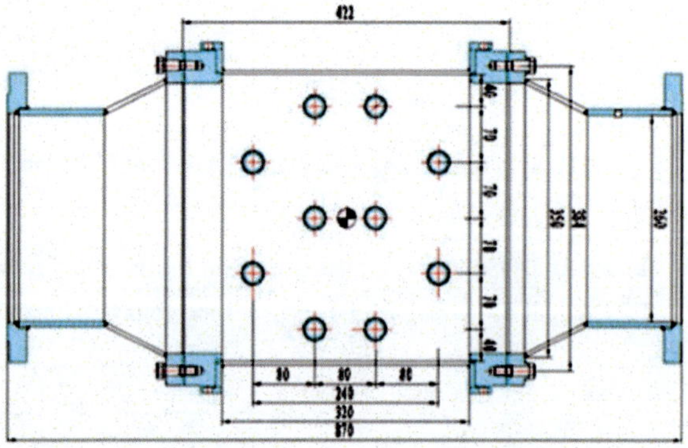

Fig. 4.13 Structure of UVC sterilizer

is composed of the lamp housing (generally pure quartz material) and the active gas and the buffer inert gas filling which is easily excited, ionized, and decomposed to produce plasma. The microwave electrodeless tube filled with Cd active gas and Ar buffer gas can stimulate 229 ~ 326 nm UVC, which can effectively meet the sterilization requirements of ballast water. The microwave non-polar ultraviolet lamp tube structure is shown in Fig. 4.14.

The advantages of the microwave cathode lamp tube UVC bacteria device are: the power conversion efficiency of the lamp driving power is high and the cost is low; the length of the lamp tube is long and the service life is long. The drawback is that independent resonators are needed, which can only work in single tube, with low flow rate and large volume of sterilizers. The microwave nonpolar tube UVC bacteria apparatus is shown in Fig. 4.15.

2. **The working principle of photocatalytic bactericide**

   There are two ways of photocatalytic sterilization:

   One is to inject O_3 through venture, the ballast water with O_3 is dissolved and then irradiated with the wavelength of 254 nm ultraviolet light, the intensifying substance, OH free radical is produced, and the ultraviolet bactericidal action is also used to kill the residual microorganisms through the synergistic action. The other method is to use a low-pressure mercury lamp with a central wavelength of 254 nm as the light source. By anodizing the TA1 titanium plate as the substrate, a fixed state TiO_2/Ti photocatalyst is prepared, and the catalyst is modified by photochemical oxidation reduction. The TiO_2 photocatalyst acts with H_2O to produce OH radicals, and also uses ultraviolet disinfection to kill microorganisms remaining after filtration through synergistic action.

   The sterilization of OH radicals first begins with the cell wall of the bacteria. The OH radical can destroy the cell wall structure, break the cell wall, and disintegrate the plasma membrane, then enter the inner cell to destroy the intima and cell components, make the cytoplasm agglomerate, lead to the spillover of the cell content, and finally the cavitation phenomenon of the cell.

   The main problems in the method of photocatalytic ozone are: (1) The rate of ozone reaction is low, the time required to kill the organisms is longer, and

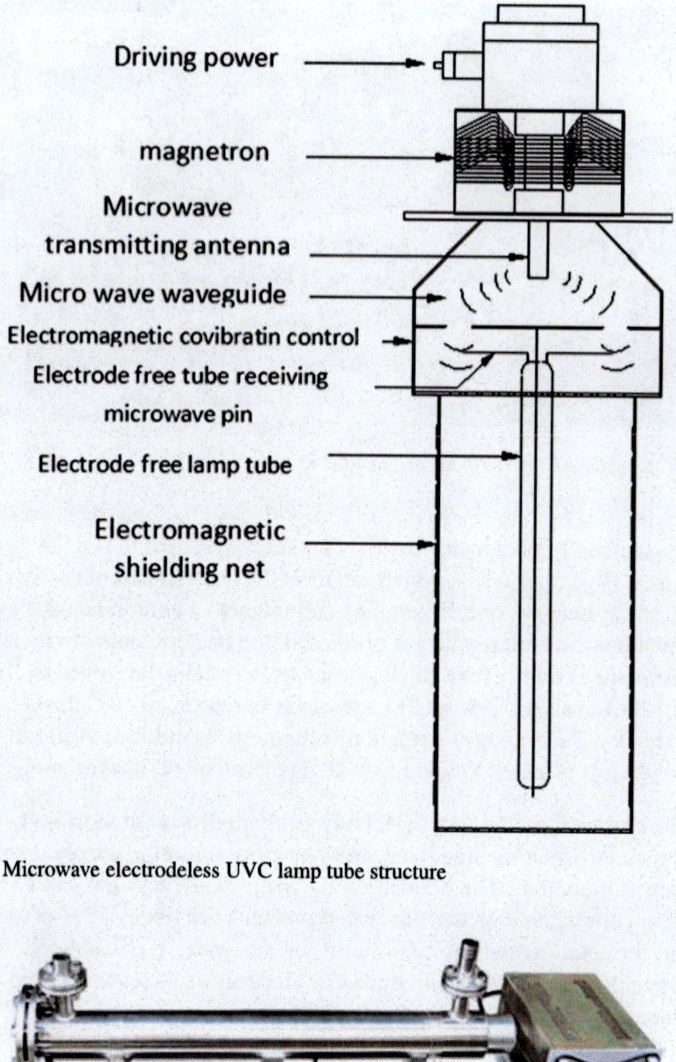

Fig. 4.14 Microwave electrodeless UVC lamp tube structure

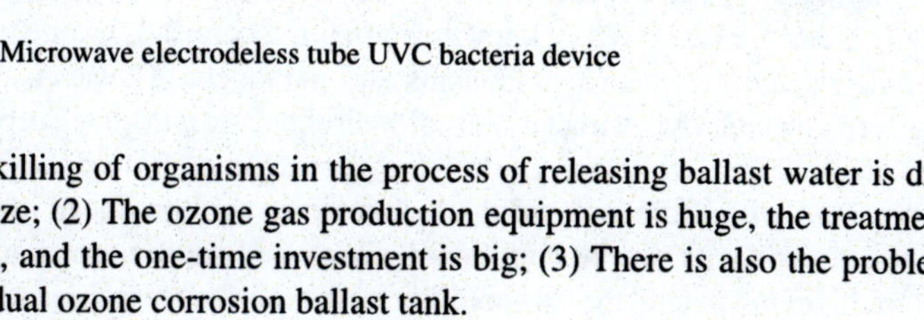

Fig. 4.15 Microwave electrodeless tube UVC bacteria device

the killing of organisms in the process of releasing ballast water is difficult to realize; (2) The ozone gas production equipment is huge, the treatment cost is high, and the one-time investment is big; (3) There is also the problem of the residual ozone corrosion ballast tank.

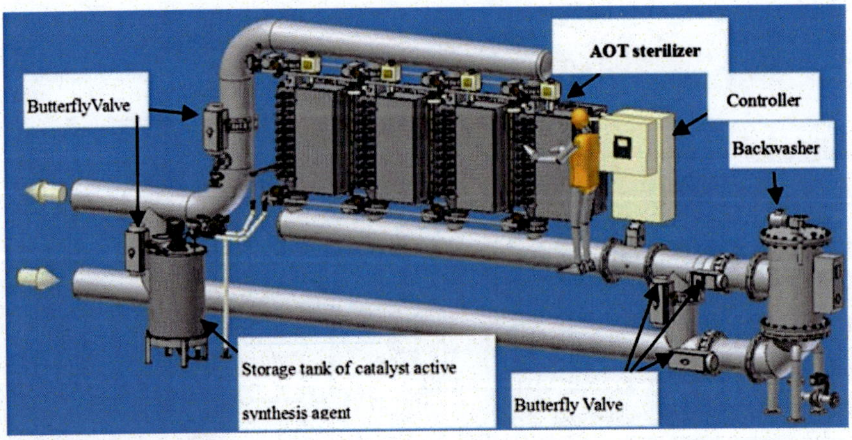

Fig. 4.16 Schematic diagram of Wallenius AOT system

The method of photocatalytic TiO_2 photocatalyst uses the OH hydroxyl radical to treat the ballast water, which can meet the D2 standard of the IMO G8 guide, but the system process is complex and requires a comprehensive agent. Every time after the ballast water is processed, the bactericidal activity is cleaned to ensure the activity of the catalyst, such as PureBalla developed by the company of Alfa Lavel in Norway. The system is based on advanced AOT (Advanced Oxidation Technology) oxidation technology and does not use any chemical reagents. It is called Wallenius AOT. The principle is shown in Fig. 4.16.

3. **The working principle of electrolysis or medicament sterilizer**
The electrolysis method uses seawater as a natural electrolyte solution with great salt content. The content of Na^+, Mg^+, Ca^+, Cl^- ion salt Cl^- ion in the main components of the sea water accounts for about 55% of the total ion, and the total content of chlorine in the sea water is 19.344 g/L, and sodium hypochlorite (NaClO) is produced by electrolysis of seawater. The electrolysis process is shown in Fig. 4.17.
In the electrolysis process in the electrolysis tank, chlorine gas (Cl_2) and two electrons (2e) are produced on the anode. On the cathode, two water molecules obtain two electrons and produce two hydroxyl groups ($2{\cdot}OH^-$) and a hydrogen molecule (H_2). Chlorine gas can dissolve in water to produce hypochloric acid (HClO) and hydrochloric acid (HCl), of which chloric acid (HClO), hypochlorite (ClO^-), and Chlorine (Cl_2), known as effective chlorine, has a strong ability to oxidize and kill aquatic organisms and pathogens. However, this method needs to consider the characteristics of hydrogen and chlorine gas produced by two kinds of gas, which are explosive and poisonous. It needs to be collected and treated with diffusion, and the hydrochloric acid produced in the process of electrolysis will also be considered, which will cause corrosion of the ship cabin. In order to avoid the two pollution of the marine environment, a moderate

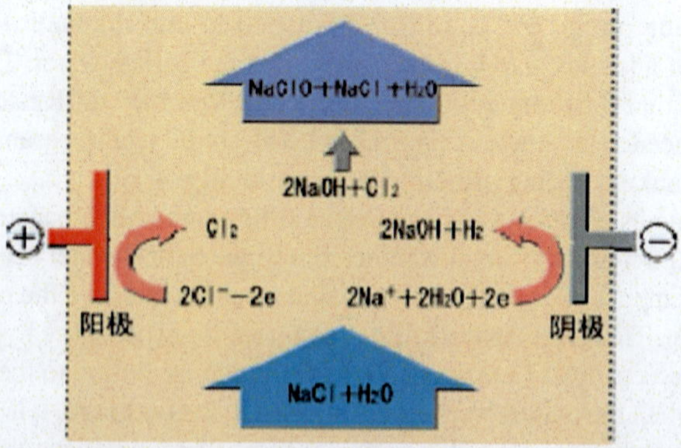

Fig. 4.17 The process of seawater in electrolysis cell

amount of neutralizer, sodium thiosulfate ($Na_2S_2O_3$), should be added to the discharge of the ballast water to the outboard. The neutralizer is usually arranged in proportion to 3 ~ 7Kg of sodium thiosulfate in the water of 1000 m, and the two kinds of harmless salts, such as sodium sulfate ($NaHSO_4$) and salt (NaCl), are synthesized by the reaction with hypochloric acid and water. Therefore, the whole operation and management system of the system is more complex and the cost of operation and maintenance is higher. Shown in Fig. 4.18 is the Qingdao double Swiss BalClor marine ballast water management system.

The reagent method is a direct use of sodium hypochlorite to configure the aqueous solution of a medicine agent. It is injected into the ballast water by

Fig. 4.18 Ballast water management system of Qingdao shuangrui ship

a metering pump, and is uniformly dispersed into the ballast water under the action of a jet pump to kill the plankton in the ballast water. The dosage of the agent is that for every 10000 m^3 of the ballast water, 100Kg sodium hypochlorite is needed to make a pharmaceutical solution with granule powder. The phytoplankton killing principle is shown in Fig. 4.19.
The same as electrolysis, because the drug method is a bactericide configured by sodium hypochlorite, it is injected into the ballast water according to certain measurement and mixed with the ballast water to achieve the purpose of killing plankton. Therefore, while killing plankton, it will produce hypochloric acid as well as electrolysis.In order to avoid secondary pollution to the marine environment, when the ballast water is discharged to the outboard, it is also necessary to add an appropriate neutralizer sodium thiosulfate in ballast water. The comprehensive treatment principle of the unloading method for ballast water is shown in Fig. 4.20.

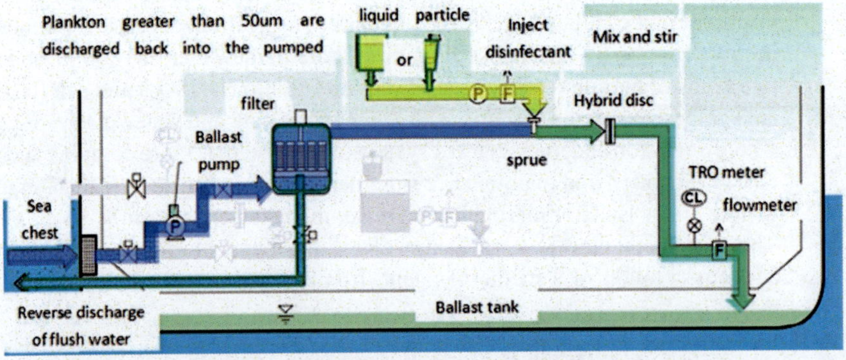

Fig. 4.19 Pharmacy method to kill plankton principle

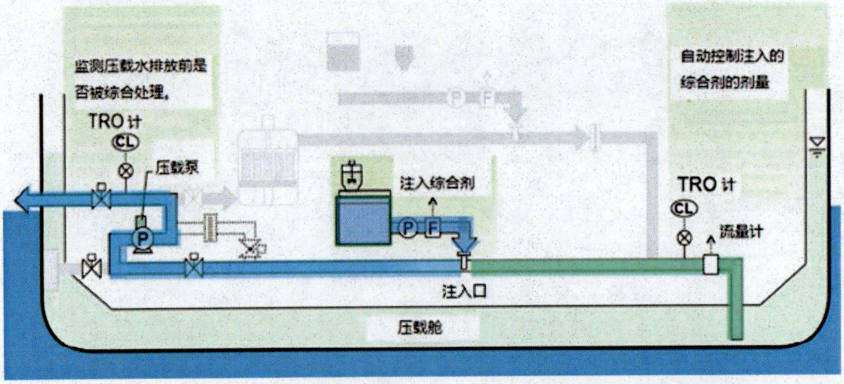

Fig. 4.20 Principle of Comprehensive Treatment of Ballast Water by Pharmacy Method

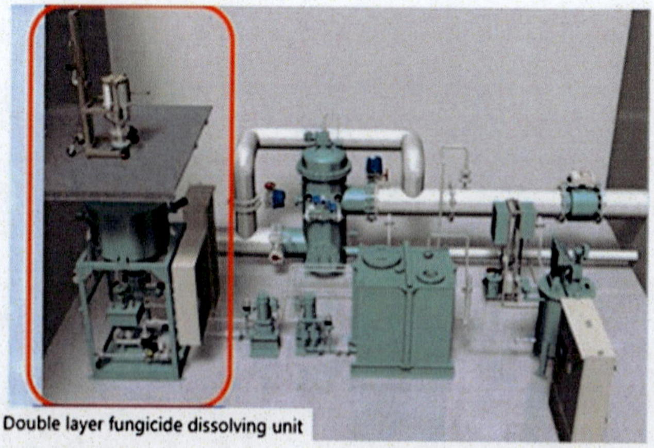

Fig. 4.21 Physical model of ballast water management system for pharmaceutical

The physical model of the medicament ballast water management system is shown in Fig. 4.21.

4. **Working principle of inert gas sterilizer**
The inert gas sterilizer uses the principle that the microorganisms are difficult to survive by leaving the oxygen, and the oxygen in the water is driven away by pure nitrogen or an inert gas to inhibit the proliferation of bacteria.Inert gas sterilizer is mainly designed according to the law of gas partial pressure (Henry's law) and total pressure law (Dalton's law). Among them: Henry's law is defined as: the solubility of a gas in water is directly proportional to the composition of the gas on the liquid surface. The definition of Dalton's law is that the total pressure of a gas mixture is equal to the sum of the partial pressure of each gas. The content of oxygen in natural atmosphere is about 20%, and the nitrogen content is about 79%. If the content of nitrogen in the airtight space increases to 95%, at this time, because there are other gases in the atmosphere, such as CO_2, rare gas and so on, the content of O_2will be less than 5%, and the content of dissolved oxygen (DO) in the water will be about 2–3%. In this case, the general microorganism cannot survive.
When the inert gas sterilizer works, the nitrogen gas is separated from the air by molecular sieve and filled into the closed ballast space of the ballast water to be treated. Because of the diffusion factor of the gas, the dissolved oxygen in the water will not be reduced rapidly. If nitrogen is filled into the water, the nitrogen partial pressure in the water will be as high as 722 mm—Hg. At this time, the nitrogen partial pressure has an absolute advantage and is bound to drive off the dissolved oxygen dissolved in the ballast water. This process takes only 10 s, and the nitrogen gas will always be in a positive pressure in the closed ballast tank and in the dominant state. Now the nitrogen partial pressure in the ballast system is much higher than the nitrogen dissolved saturation of the 20.6 mg/L in

Fig. 4.22 Inert gas sterilization system

the ballast water. Therefore, the supersaturated nitrogen part of the ballast water will escape from the surface of the water and will enter into the space above the surface of the water.Because of this, it is also called the supersaturation nitrogen treatment method for the ship's ballast water management system. The inert gas sterilization system is shown in Fig. 4.22.

The ballast water inert gas sterilization system is overly washed into the ballast water, which makes the dissolved oxygen in the ballast water removed, and makes the solubility of nitrogen in the seawater saturated at 20.6 mg/L. In order to maintain the saturation of nitrogen in the ballast water and remove the dissolved oxygen in the seawater, nitrogen needs to be continuously filled into the seawater. The high nitrogen partial pressure is maintained on the surface of the ballast water. Because the ballast tank is connected to the ventilation cap on the ship deck through the air permeable pipeline in the cabin, the air in the cabin is discharged through the air permeable pipe when the ship is injected into the ballast tank, and the application of the ballast water inert gas sterilization system must be on the ship's armor. The breathing valve on the board can be modified to smooth the air from the ballast tank to outboard when the ballast waster is injected into the ballast tank. At the same time, the nitrogen in the ballast tank can be kept in positive pressure after the inert gas sterilization system works, to ensure the nitrogen saturation in the ballast water and to kill the plankton. The structure of the new breathing valve is shown in Fig. 4.23.

The new type of breathing valve consists of the air pipe flange, the lower valve disc, the floating ball, the upper valve disc, the air outlet and the air inlet. When the ballast water is injected into the ballast bank, the gas pressure in the cabin increases with the volume of the injected ballast water, and the air in the cabin must be compressed along the air pipe to reach the breathing valve. At this point the gas is split into two ways. One way along the air inlet upwards, acting

Fig. 4.23 Structure of novel ventilation cap

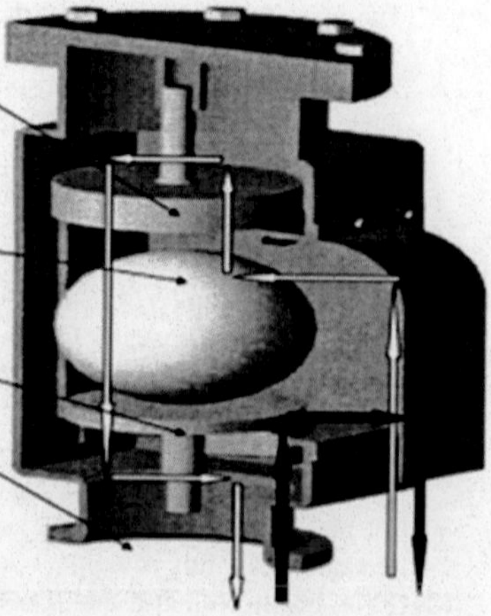

directly above the upper valve disc, so that the upper valve disc is pressed against the upper valve disc seat, and the other is acting below the lower valve disc, when the gas pressure reaches a certain value, the positive air pressure Will overcome the weight of the lower valve disc and the float ball to jack up the lower valve disc, so that the cabin gas is discharged.When the ballast water is unloaded, at this time, under the action of the unloading pump, the ballast water is discharged outboard, and as the ballast water is discharged, the pressure of the ballast tank liquid level gradually changes from a positive pressure to a negative pressure. When the lower valve disc of the breathing valve is adsorbed on the lower valve disc seat under the action of the negative pressure, since the external air pressure is greater than the pressure in the cabin, atmospheric pressure acts on the bottom of the upper valve disc in the breathing valve. The valve disc is opened, so that the external atmosphere enters the ballast tank through the air inlet of the breathing valve, thereby ensuring smooth unloading.

Since the ballast water inert gas sterilization system uses the filtration + membrane separation + nitrogen flooding technology, the organic matter in the treated ballast water is rarely dead or alive. Therefore, it does not promote the occurrence of anaerobic bacteria.

References

1. Ai-wei, Wang. 2008. Development and Theoretical Research of Sucking Backwash Filter. Beijing University of Chemical Technology.
2. Ai-wei, Wang, Zhu Hai-jun and Qian Cai-fu et al. 2007. Back flush filter structure characteristics and application prospects. *Petrochemical Equipment*.
3. Fei, Liu, Tan Hui and Yang Jian et al. 2007. Practicability of Hydro cyclone-ultraviolet treatmentof ship's ballast water. *Chinese Navigation* (4): 100–103.

4. Yue-juan, Yan. 2014. Research on Solid-liquid Two-phase Flow Characteristics in Downstream Swirling Sand Remover. Daqing: Northeast Petroleum University, Ph.D. Dissertation, 15–54.
5. Xian-zhen, Meng. 2013. Study on Marine Biological Technology of Ship Ballast Water. Zhenjiang: Masters of Science Dissertation of Jiangsu University of Science and Technology, 19–22.
6. Bajram, Z., H. Mark and A. J. Carroll. 2006. Studies of a novel sensor for assessing the spatial distribution of cavitation activity within ultrasonic cleaning vessels. *Ultrasonics* 44(1): 73–82.
7. Ent Ezari, M et al. 1997. The effect of frequency on sonochemical reactions III :dissociation of carbon disulfide. *Ultrasonics S Onochemistry* (4): 49–54.
8. Wang Hai-ou, Hu., Wang Jian-nan Zhi-chao et al. 2011. Effects of connection mode between transducer and tank on ultrasonic field distribution. *Transactions of the Chinese Society of Agricultural Engineering* 27(2): 26–30.
9. Zhen-hua, Ding, Fan Jian-zhong et al. 2000. *Ultraviolet Radiation Biology and Medicine*, 20–24. Beijing: People's Military Medical Press.
10. Yu-jun, Chen, Guo Ping and Liang Chao-xiong. 2007. The necessities and key techniques for the ballasthandling equipment of oceangoing ships. *Chuanhai Engineering* 36(3):30–31.
11. Ming-hong, Wu., and Bao Bo-rong. 2002. *Application of Radiation Technology in Environmental Protection*, 51–60. Beijing: Chemical Industry Press.
12. Klan, Petr, Jaromir Literak and Milan Hajek. 1999. The electrodeless discharge lamp:a prospective tool for photochemistry. *Journal of Photochemistry and Photobiology. A: Chemistry* (128): 145–149.
13. Muller, Pavel, Petr Klan and Vladimir Cirkva. 2003. The electrodeless discharge lamp:a prospective tool for photochemistry Part 4: Temperature and envelope material dependent emission characteristics. *Journal of Photochemistry and Photobiology. A: Chemistry* (158): 1–5.

Chapter 5
Design Method of Key Components

To efficiently kill plankton in ship's ballast water, current methods use a combination of filtration and other related sterilization technologies, which reached the goal of transient sterilization in large flow ballast water over-current, to ensure the safety of ship loading and navigation. As a result, when designing a ballast water management system, it is necessary to master the design methods of the various types of equipment that make up the system.

5.1 Design of Filter Equipment

The filter is an important part in the ballast water management system. It ensures that the sediment and the plankton in the sea water are blocked before the germicide. Thus, the particles of sediment and plankton are less than 50 μm in the water quality of the germicide, and the turbidity of the water is cleared and the bactericidal efficiency of the sterilizer is improved. At present, the main filters used in ballast water system are two types: backwash filter and whirling filter.

5.1.1 Design of Backwash Filter

Filtration process: raw water flows from the filter inlet and flows through the filter from inside to outside, and the water flows out from the outlet. Impurities accumulate in the inner surface of the filter to form filter cake, resulting in differential pressure inside and outside the filter. When the pressure difference reaches the set value of the differential pressure switch, the self-cleaning will be performed [1].

Self-cleaning process: cleaning relies on a sucking scanner that moves along the inner surface of the screen. The scanner is also known as a suction device. It has a hollow structure, and a number of suction pipes are uniformly distributed at a certain

N. Chen et al., *Development and Implementation of Ship BWMS*,
https://doi.org/10.1007/978-981-33-6865-1_5

Table 5.1 Basic parameters of self-cleaning filter

Filter flow rate Q	300 m^3/h	Design length L	1600 mm
Material	316 L stainless steel	inlet and outlet diameter d_2	209 mm
Filter area	5500 cm^2	inlet and outlet velocity v	2.43 m/s
Inner diameter of cylinder d	450 mm	Mesh	Composite sintered stainless steel wire mesh
Total height h	2050 mm		

distance along its axis. The cleaner is connected with the blowdown valve. When the blowdown valve is opened, the pressure difference between the water pressure inside the filter and the external atmospheric pressure makes the suction pipe build a strong suction force. The high-speed water flows from the outside of the filter in the reverse direction to flush the impurities into the suction device and discharge through the drain valve.The suction device is driven by a two-way motor with a screw unit and is screwed at a fixed speed so that multiple nozzles can be absorbed throughout the screen surface. The system keeps flowing throughout the anti-cleaning process.The entire operation of the filter is controlled by a PLC [2].

The basic structure of the suction type backwash filter includes: cylinder, head, suction device, filter screen, screw shaft, driving device, inlet and outlet, sewage valve, and so on. Taking the design of a suction backwash filter with 300 m^3/h processing capacity as an example, the specific performance parameters are shown in Table 5.1.

The working time of a backwash filter is controlled by continuous differential pressure and can be continuously washed. Full automatic intelligent control and fault alarm are used.

The assembly drawing of the suction type self-cleaning filter is shown in Fig. 5.1.

1. **Selection and design of filter screen**

To determine the parameters of the filter, first, the hydraulic design of the filter is needed. The hydraulic design of the filter includes filtration speed design and filtration capacity- flow calculation.

(1) **Determination of filtration velocity**

The speed at which water flows through the filter is called the filtration speed. When the filtration area is constant, the larger the filtration speed, the stronger the filtration capacity (flow rate), but some of the sewage in the treated wastewater is easy to squeeze the filter into the clean water side pipeline, so that the filtered water quality does not meet the requirements for use. At the same time, it will produce larger head loss and increase the pressure difference between the two sides of the filter. As a result, it may damage the filter. If the filtration rate is too small, in order to achieve a certain filtration capacity, a larger filter mesh area will be needed, which

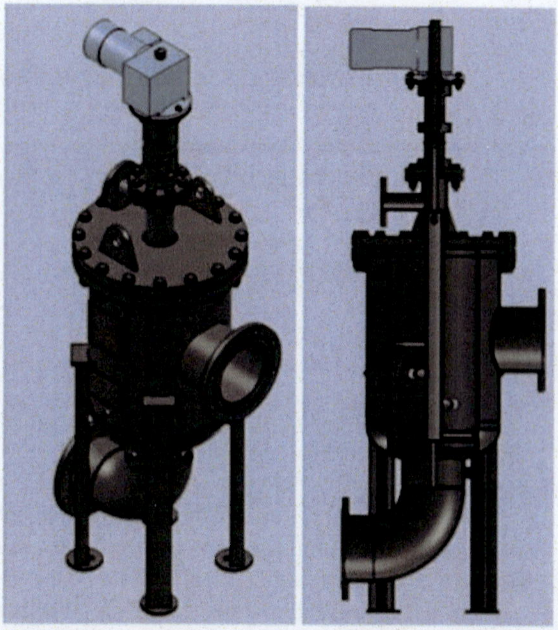

Fig. 5.1 Self-cleaning suction filter

will increase the volume of the filter and increase the cost. Therefore, filtering speed is one of the important parameters for designing filters. The filtration speed of the backwash filter is usually selected between 0.5 m/s and 3 m/s, and the velocity of water flow in the ballast water pipe is generally between 2.5 m/s and 3 m/s, and the filter speed is less than the flow velocity of the ballast water in the pipe. Based on the above considerations, it is advisable to select the filtration speed v = 1.5 m/s for the general purpose.

(2) **Determination of an effective filtration area**

The relationship between flow rate, effective filtration area, and filtration rate is as follows:

$$Q = 3.6 \times 10^3 \times f \times A \times v \tag{5.1}$$

where

Q the design flow rate of the filter (m^3/h);
A the actual filter area of the filter (m^2);
v filtering speed (m/s);
f the net area coefficient of the filter.

The flow rate of the filter is 300 m^3/h and the filter speed is 1.5 m/s. To design a filter that meets the requirements of the ballast water filtration, the net area coefficient of the filter must be determined.

Fig. 5.2 Cross-section structure of sintered metal mesh

Considering the corrosiveness of seawater, the filter type generally selected is composite sintered 316L stainless steel wire mesh, as shown in Fig. 5.2.The sintered stainless steel mesh microporous filter material is a multi-layer mesh layer and is prepared by vacuum diffusion sintering filtration technology. The filter material has the advantages of uniform pore size, good overall rigidity of the filter material, extensive engineering application, and easy backwashing regeneration.

From the cross-section structure of the filter material shown in Fig. 5.2, it can be seen that the precision stainless steel wire mesh used in the filter layer is the key working layer of the sintered stainless steel screen filter material. The typical knitting methods of precision stainless steel screen are two kinds of plain weave and twill weave. Among them, the filter material produced by the plain weave method has the advantages of high opening rate and easy to calculate the filtering accuracy according to the mesh size, so it is preferred by most designers. Table 5.2 shows the basic parameter of the plain woven stainless steel silk dense filter in the market. It can be seen from the table that the effective area rate of the filter net of the aperture of

Table 5.2 Plain weave microgroove filter gauze

Model MPW	Wire diameter Latitudes (mm)	Basic pore size reference value (μm)	Effective cross section rate (%)	Quality of unit area network(kg·m^{-2})	Net thickness (mm)	The equivalent of the British system by the latitude
160	0.20 × 0.14	60	12.5	1.40	0.48	40.6 × 200
180	0.16 × 0.112	56	15.0	1.09	0.384	45.4 × 249
200	0.14 × 0.10	50	14.6	0.96	0.34	51.3 × 279
240	0.20 × 0.10	45	12.8	0.99	0.34	57.7 × 279
260	0.125 × 0.09	36	10.8	0.90	0.305	67.7 × 310

50 m is 14.6%, that is, f = 0.146. Considering that the filter medium is the effective filter area of the multilayer sintered stainless steel network, the area of the filter area will be reduced, so the smaller net area is properly taken when the filter area is calculated. Number. The net area coefficient f = 0.1 of the filter screen is selected.

The actual filtration area of 316L stainless steel sintering filter net can be calculated as follows:

$$A = \frac{Q}{3.6 \times 10^3 \times f \times v} = 555556\ \text{mm}^2 \tag{5.2}$$

(3) Determination of the filter screen size

The filter screen of the suction type backwash filter is cylindrical, and the filtration area refers to the inner surface area of the filter screen.

$$A = \pi DL \tag{5.3}$$

where:

A the actual filter area of the filter (m^2);
D the inner diameter of the filter element (m);
L the effective length of the filter element (m).

From the perspective of stress and stability, it is advisable to consider the slender λ (L/D) of components in general from 1 to 5. Considering the limitation of the space of the engine room, 2 is usually chosen.

$$\lambda = L/D = 2 \tag{5.4}$$

Based on equatons (5.2), (5.3), and (5.4), the followingparameters can be obtained: D = 0.297 m = 297 mm and L = 0.594 m = 594 mm, which can be rounded to: D = 300 mm and L = 600 mm.

2. **Design of filter barrel**

The cylinder and head of the backwash filter belong to the thin-wall revolving shell, and must be designed according to the design specification of the pressure vessel.

(1) **Determination of the technical parameters of the filter barrel**

The design parameters of pressure vessel mainly include design pressure, design temperature, cylinder material etc. Working pressure p_w = 10 bar = 1 MPa, design pressure p = 1.1p_w = 1.1 MPa, and calculation pressure p_c = p=1.1 MPa; design temperature t = 80 °C; considering the corrosive nature of seawater, the cylinder material is 316L stainless steel.

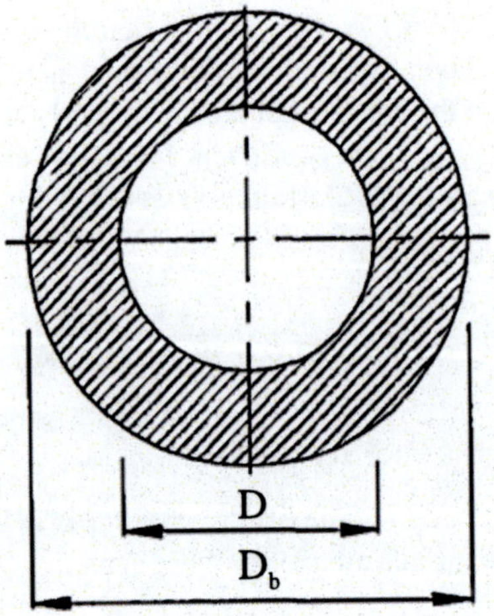

Fig. 5.3 Diameter of tube body

(2) Cylinder diameter calculation

The area between the filter screen and the shell should meet the requirement that the velocity of the flow through the area should be equal to or less than the longitudinal flow velocity inside the filter element, that is, the V cylinder is less than V yuan. This means that the A cylinder is more than A yuan. As shown in Fig. 5.3, the total area of the filter cartridge is:

$$A_b = A_c + A_r \geqslant 2A_r \tag{5.5}$$

And the $A_b = \pi\ D_b\ /4$, $A_c = \pi\ D/4$.Substitute the two equations into formula (5.5), we can obtain the following:

$$D_b \geqslant \sqrt{2}D \tag{5.6}$$

$D_b = 1.5D = 1.5 \times 300 = 450$ mm.

The inner diameter of the cylinder of the pressure vessel should conform to the nominal diameter of the vessel. Since the cylinder of the container is to be matched with the flange and the support, the cylinder is selected by the nominal diameter series flange and the support standard, and is customized according to the nominal diameter series of the container.

For the cylinder welded by steel plate, the inner diameter of the cylinder is specified as the nominal diameter, and its series size is shown in Table 5.3. The diameter of the cylinder body $D_1 = D_b = 450$ mm can be determined.

Table 5.3 Diameter series of pressure vessel (mm)

300	350	400	450	500	550	600	650
700	750	800	850	900	950	1000	1100
1200	1300	1400	1500	1600	1700	1800	1900
2000	2100	2200	2300	2400	2500	2600	2700
2800	2900	3000					

(3) **Packing seal design**

The packing seal is designed according to the seal design manual and related standards.

① **The determination of the basic parameters of the seal**

The transmission shaft of the filter is made of intermittent low speed spiral motion, so the polytetrafluoroethylene woven packing seal is adopted; the medium pressure is p = 1 MPa, the temperature is 80, the transmission shaft diameter is d = 20 mm, the transmission shaft speed is n = 18 r/m in, and the circumferential velocity of the shaft is v = π nd = 0.0188 m/s.

The shaft is designed to rotate and reciprocate, and the maximum pressure of the medium is 1 MPa. Therefore, the number of packing rings is n = 5; according to the size of the journal, the width of the section edge of the filler is selected as S = 6 mm, so the filling height is H = ns = 30 mm.

② **Calculation of the diameter of the cap bolt**

The force Q_1 required to compress the packing is calculated as follows.

$$Q_1 = 78.5(D^2 - d^2)y \tag{5.7}$$

where:

- *y* compaction pressure (MPa), high quality asbestos packing y ≈ 4; jute filler y ≈ 2; flexible graphite filler y ≈ 2.5;
- *d* axis diameter (cm);
- *D* the diameter of the inner wall of the packing box (cm).

As an example, let us take y = 4 MPa, d = 2 cm, and D = d+2 s = 3.2 cm and substitute them into Eq. (5.7) above. It will yield Q_1 = 1959 N.

The force required to seal the stuffing box is Q_2, which is calculated as follows.

$$Q_2 = 235.6(D^2 - d^2)p \tag{5.8}$$

Substitute $p = 1$ MPa, $d = 2$ cm, and $D = 3.2$ mm into Eq. (5.8), we will get Q_2 = 1941 N.

The thread diameter of the gland bolt is determined by bolt load $Q = \{Q_1, Q_2\}\text{max}$ = 1959 N. That is:

$$Q_{\max} \leq 25\pi d_0^2 Z\sigma_p \tag{5.9}$$

where:

Z the number of bolts is usually 2, 3 or 4.
σ_p Bolt permissible stress (MPa);
d_0 The bolt screw path (cm).

Formula (5.9) can be transformed into:

$$d_0 \geq \sqrt{\frac{Q_{\max}}{25\pi Z\sigma_p}} \tag{5.10}$$

The bolt material is selected as Q235 and the performance grade is 6.8. The yield limit of the material is $\sigma_s = 480$ MPa, the safety factor is $S = 12.5$, $\sigma_p = \frac{\sigma_s}{S} = 38.4$ MPa, and the $Z = 4$ substitute those parameters into Eq. (5.10), we will get $d_0 > 4$ mm. According to the standard of coarse tooth common thread (GB/T196-2003), the nominal diameter of thread is $d = 6$ mm (thread diameter $d_0 = 4.918 > 4$ mm).

③ Calculation of friction power

Friction F_m between the filler and the rotating shaft can be calculated as follows.

$$F_m = 100\pi dHq\mu \tag{5.11}$$

where:

q the side pressure of the packing (MPa), usually $q = K\frac{Q_{\max}}{25\pi(D^2-d^2)}$
K lateral pressure coefficient. For oil immersed natural fiber, $K = 0.6–0.8$, asbestos $K = 0.8$~0.9, flexible graphite braided packing $K = 0.9–1$;
μ The friction coefficient between filler and rotating shaft is $\mu = 0.08$~0.25;
d Axis diameter (cm);
H The height of the packing (cm).

In this case, let us set $K = 1.0$, $\mu = 0.25$, H = 3 cm, and = 2 cm which gives $q = K\frac{Q_{\max}}{25\pi(D^2-d^2)} = 4$ MPa.Subtituting it into Eq. (5.11) gives $F_m = 1884$ N.

In the whole filling height of the stuffing box, the lateral pressure distribution is uneven, and the pressure from the packing gland to the bottom of the packing is gradually decreasing. Therefore, the friction power is approximately calculated according to the following formula:

$$P = \frac{F_m v}{1000} \tag{5.12}$$

where:

μ circumferential velocity (m/s);

d Axis diameter (m);
n The rotational speed of the axis (r/s).

As an example, let us set $n = 0.3$ r/s and $v = \pi dn = 0.0188$ m/s and substitute them into Eq. (5.12) to obtain $P = 0.035$ kW.

3. **Design of transmission parts**

(1) **Design and selection of motor and reducer**

The suction device of the backwash filter requires a spiral motion at a fixed rotational speed so that a plurality of nozzles sweep across the entire inner surface of the screen. Therefore, a two-way drive motor and shifting mechanism with a screw device are required.

The wear power of the packing seal of the cleaner is 0.035 kW, the power consumed by the fluid in the operation of the cleaner and the friction power of the other supporting parts at the operation of the absorber is very small. Therefore, the output power of the screw rod does not exceed 0.05 kW, and the safety factor of the screw is 1.5 times that of the abnormal condition. The output power of the screw rod does not exceed 0.075 kW. The mechanical transmission efficiency of the screw rod is 0.3, and the mechanical transmission efficiency of the initial decelerator is 0.95. Therefore, the required motor power is $P_1 = 0.26$ kW. Choose small power three-phase asynchronous motor YS7124 with power $P_1 = 0.37$ kW, speed $n_1 = 1400$ r/min, and rated torque $T_1 = 2.4$ N m.

The required reducer transmission ratio is $i = \frac{1400}{18} = 77.78$, the reducer output power is $P_2 = P_1\eta = 0.37 \times 0.95 = 0.35$ kW, the output speed is $n_2 = 18$ r/min, and the output torque is $T_2 = 9550\frac{P_2}{n_2} = 186$ N m.

Select K series combined helical gear-bevel gear reducer produced by a factorywith the model number KAF47YS7124, whose performance parameters are as follows: power $P = 0.37$ kW, output speed $n = 18$ r/min, output torque $T = 180$ N·M, and transmission ratio $i = 75.2$.

(2) **Design and verification of wire rod**

The motor drives the screw shaft to rotate through the key connection, and the screw shaft is connected through the coupling and the dirt suction device. In order to reduce the filter height, the size of the screw shaft should be as small as possible. The screw structure is shown in Fig. 5.4.

① **Upper connecting shaft check**

The upper shaft segment is connected with the reducer, and the shaft material is 316L stainless steel. In order to match the motor purchased, the diameter of the shaft is 35 mm. Calculate the strength of the shaft according to the torsion of the shaft.

$$d_1 \geq \sqrt[3]{\frac{9550000 P_1}{0.2n[\tau]}} \tag{5.13}$$

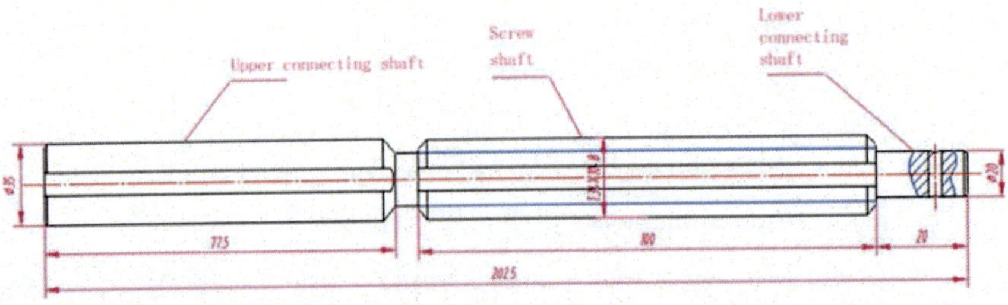

Fig. 5.4 Structure of screw

where:

d_1 the diameter of the axis (mm);
P_1 The power of the axle transfer (kW);
n The rotational speed of the axis (r/min);
$[\tau]$ It is allowed to use torsional shear stress (MPa);

Motor $P = 0.37$ kW, the drive efficiency of the bevel gear reducer is 0.95, so the transmission power of the shaft is $P_1 = 0.37 \times 0.95 = 0.35kW$; speed of the shaft $n = 18\text{r/min}$; the relationship between the torsional shear stress and the tensile allowable stress is $[\tau] = (0.5 \sim 0.6)[\sigma]$ $[\tau] = 0.5[\sigma] = 0.5 \times 118 = 59$ MPa. Substituting those parameters into the Eq. (5.13) gives $d_1 \geq 25.1$ mm. For the shaft less than 100 mm, there is a keyway, the axle neck increases by 5%~7%, $1.07d_1 = 26.9$ mm < 35 mm and the strength of the shaft is satisfied.

② Screw check

The parameters of the shaft shaft are as follows: the height of the nut $H = 50$ mm, the screw pitch $P = 10$ mm, the guide range $L = 10$ mm, the tooth angle $\alpha = 30°$, the number of thread working rings $z = \frac{H}{P} = 5$, the diameter of the screw thread $d_2 = 29$ mm, the screw thread working height $h = 0.5P = 5$ mm, the screw thread diameter $d_1 = 23$ mm, the screw thread outer diameter $d = 34$ mm, the length of the screw is greater than the sum of the stroke of the suction device and twice of the length of the nut, taking a length of 200 mm; the inner diameter of the nut $D = 35$ mm.

Both screw and nut materials are 316L, yield limit\sigma$a_s = 480$ MPa, allowable stress $[\sigma] = 118$ MPa, allowable bending stress $[\sigma_b] = (1.0 \sim 1.2)[\sigma] = (118 \sim 142)$ MPa and allowable shear stress $[\tau] = 0.6[\sigma] = 70.8$ MPa.

When the sliding screw works, it mainly bears the role of torque and axial tension (or pressure), and at the same time, there is a larger relative slip between the screw threads of the screw and nut. Its failure form is mainly threaded wear.

I. **Self locking checking**

$$\varphi = \arctan\frac{L}{\pi d_2} = \arctan\frac{10}{3.14 \times 29} = 6.27° \tag{5.14}$$

$$\rho = \arctan\frac{f}{\cos\frac{\alpha}{2}} = \arctan\frac{0.14}{\cos 15^\circ} = 8.25^\circ \tag{5.15}$$

Because of $\varphi < \rho$, it is self locking and reliable. where:

φ thread angle;
ρ equivalent friction angle;
f The friction coefficient of the screw pair and the friction coefficient between steel and steel $f = 0.11 \sim 0.17$. Take its average value of 0.14.

II. Calculation of transmission efficiency

$$\eta = (0.95 \sim 0.99)\frac{\tan\varphi}{\tan(\varphi \pm \rho)} \tag{5.16}$$

(0.95–0.99) is the bearing efficiency, which is determined by the bearing form, and the sliding bearing takes a small value. When the axial load is opposite to the direction of movement, the " + " number is taken. As an example, do the following:

$$\eta = 0.95 \times \frac{\tan 6.27^\circ}{\tan(6.27^\circ + 8.25^\circ)} = 40.3\% \tag{5.17}$$

III. Calculation of wear resistance

The force of the component itself is smaller than that of the rod under the pressure in the liquid, and its axial force Q is negligible. The transmission power of the screw rod does not exceed 0.05 kW, and the safety factor of the 1.5 times of the abnormal condition is obtained. The transmission power of the screw rod is 0.075 kW. At this point, torque is $T_0 = \frac{9550P}{n} = 39.8\ N \cdot m$ and the torque of the wire rod $T = \frac{T_0}{\eta} = 98.8\ N \cdot m$.

Axial force can be calculated from torque $T = F\tan(\varphi + \rho)\frac{d_2}{2}$ as follows:

$$F = \frac{T}{\tan(\varphi + \rho)\frac{d_2}{2}} = \frac{98800}{\tan(6.27^\circ + 8.25^\circ) \times \frac{29}{2}} = 26.3\text{KN} \tag{5.18}$$

$$p = \frac{F}{A} = \frac{F}{\pi d_2 h z} = \frac{26300}{3.14 \times 29 \times 5 \times 5} = 11.6\text{MPa} < 12\text{MPa} \tag{5.19}$$

In this case, the thread transmission belongs to low speed sliding. At this time, the allowable stress of steel and steel is 7.5–13 MPa, and the allowable stress 12 MPa is taken to meet the wear resistance conditions. d_2 meets the requirements.

The height of the nut is $H = \phi d_2$ 1.2–2.5 for the whole nut and 2.5–3.5, $H = \phi d_2 = 34.8 \sim 72.5$mm for the partial nut. Between 34.8 and 72.5, the requirements are met.

IV. Calculation of screw strength

The screw is subjected to axial force F and torque T, according to the fourth strength theory:

$$\sigma = \sqrt{\left(\frac{4F}{\pi d_1^2}\right)^2 + 3\left(\frac{T}{0.2d_1^3}\right)^2} = \sqrt{\left(\frac{4 \times 26.3 \times 10^3}{3.14 \times 23^2}\right)^2 + 3\left(\frac{98.8 \times 10^3}{0.2 \times 23^3}\right)^2} = 95\text{MPa} \quad (5.20)$$

So $\sigma = 95MPa < [\sigma] = 118MPa$, the strength of the rod is satisfied

V. Strength calculation of screw thread teeth

Whorl tooth bottom width $b = 0.65P = 6.5$ mm

The shear strength of the dangerous section of the root of the thread:

$$\tau = \frac{F}{\pi d_1 bz} = \frac{26300}{3.14 \times 23 \times 6.5 \times 5} = 11.2\text{MPa} < [\tau] = 70.8\ \text{MPa} \quad (5.21)$$

The bending strength condition of the dangerous section of the root of the thread:

$$\sigma_b = \frac{3Fh}{\pi d_1 b^2 z} = \frac{3 \times 26300 \times 5}{3.14 \times 23 \times 6.5^2 \times 5} = 25.9\ \text{MPa} < [\sigma_b] = (118 \sim 142)\ \text{MPa} \quad (5.22)$$

According to the formula (5.21) and (5.22), the strength of screw thread teeth meets the requirements.

VI. Strength calculation of screw thread of nut

The shear strength of the dangerous section of the root of the thread:

$$\tau = \frac{F}{\pi dbz} = \frac{26300}{3.14 \times 34 \times 6.5 \times 5} = 7.8\ \text{MPa} < [\tau] = 70.8\ \text{MPa} \quad (5.23)$$

The bending strength condition of the dangerous section of the root of the thread:

$$\sigma_b = \frac{3Fh}{\pi d b^2 z} = \frac{3 \times 26300 \times 5}{3.14 \times 34 \times 6.5^2 \times 5} = 17.5\ \text{MPa} < [\sigma_b] = (118 \sim 142)\ \text{MPa} \quad (5.24)$$

VII. Calculation of the stability of the screw

The flexibility of the screw:

$$\lambda = \frac{\mu l}{i} \quad (5.25)$$

where:

λ the flexibility of the screw;

μ The length coefficient of screw is related to the end structure of screw.

i The inertial radius (mm) of the dangerous cross section of the screw.
l The working length of the screw (mm).

Screw at both ends of a screw $\mu = 1$, $l = 200$ mm , $i = \frac{d_1}{4} = 5.75$ mm, so $\lambda = \frac{\mu l}{i} = 34.8$. Because of $\lambda = \frac{\mu l}{i} = 34.8 < 90$, for 316L the critical load of the screw is:

$$F_c = \frac{340}{1 + 0.0002\lambda^2}\frac{\pi d_1^2}{4} = 113.7kN \tag{5.26}$$

Since $\frac{F_c}{F} = 4.3 > 4$, it meets the requirements.

4. **Calculation of pressure drop of the filter**

The filtration process of backwash filter is divided into two steps:

1. Screen filtration. At the initial stage of filtration, the particles in the water larger than the pore size of the filter screen are trapped by the filter screen and remain on the surface of the filter screen. At this point, the filter acts as a sieve.

2. Cake filtration. As the filtration is carried out, the solid particles suspended in the fluid are continuously deposited in the pores of the medium or on the surface of the medium to form a layer of filter cake, which is filtered through the deposited solid particle layer. This process continues until the pressure drop on both sides of the cake exceeds the economic or technical allowable maximum value, or fills the effective gap. Therefore, the pressure drop of the filter mainly includes two parts: filter pressure drop and cake pressure drop.

(1) **Filter pressure drop**

When a clean liquid passes through a porous media layer, the relationship between the volume velocity and the pressure drop is proposed by the Frenchman Darcy (Darcy) in 1856, so it is called Darcy's law. If the laminar flow of fluid in the medium is dominant, the flow can be described by Darcy equation [3]:

$$\frac{1}{A}\frac{dV}{dt} = u = \frac{B\Delta P}{\mu L} \tag{5.27}$$

where:

A the total surface area of the filter medium (m^2);
V Volume flow (m^3/s);
$\frac{dV}{dt}$ the average linear velocity (m/s) of fluid through the filtration medium;
μ fluid viscosity (Pa. S);
B the specific permeability (m^3/s) of the filter medium (filter screen);
ΔP The total pressure drop (Pa) of the fluid through the filter medium;
L The thickness of the filter medium (m/s).

The relationship between the specific permeability B of the filter and its structural geometry can be described by the Kozani Kaman equation [4].

$$B = \frac{\mu u L}{\Delta P} = \frac{1}{K_0 S_0^2} \frac{\varepsilon^3}{(1-\varepsilon)^2} \tag{5.28}$$

where:

ε porosity of filter media;
S_0 Filter medium specific surface;
K_0 The Kozani coefficient.

The filter screen used in this case is composite sintered wire mesh, and the work layer is plain weave metal mesh. The effective interface rate of the selected plain metal mesh is 14.6%, that is, the porosity of the screen $\varepsilon = 14.6\%$.

The longitude and latitude line of the filter can be regarded as a smooth cylinder d with a diameter. S_0 can be defined as the surface area per unit volume.

$$S_0 = \frac{S}{V} = \frac{\pi d \times 1}{\pi \left(\frac{d}{2}\right)^2 \times 1} = \frac{4}{d} \tag{5.29}$$

Substituting S_0 into Eq. (5.15) gives us the following:

$$B = \frac{d^2}{16 K_0} \frac{\varepsilon^3}{(1-\varepsilon)^2} \tag{5.30}$$

The relationship between S_0 and ε :

$$K_0 = \frac{4\varepsilon^3}{(1-\varepsilon)^{0.5}} [1 + 56(1-\varepsilon)^3] \tag{5.31}$$

Substitutingequations (5.17) and (5.18) into Eq. (5.15) gives us the following:

$$\Delta P = \frac{64 \mu u L}{d^2} (1-\varepsilon)^{1.5} [1 + 56(1-\varepsilon)^3] \tag{5.32}$$

Equivalent area cylinder diameter $d = 2 \times \sqrt{\frac{(140+100) \times 340/2}{\pi}} = 228\ \mu\text{m}$

Filtration speed $u = \frac{Q}{A\varepsilon} = \frac{300/3600}{0.56 \times 0.146} = 1.02\ \text{m/s}$

The pressure difference ($\Delta P_{net} = 0.012\text{MPa}$) between the inside and outside surfaces of the filter screen can be obtained via Eq. (5.32) using the following values: $d = 228\ \mu\text{m}$, $\mu = 1.005\ \text{mPa} \cdot \text{s}$, $u = 1.02\ \text{m/s}$, $\varepsilon = 0.146$, $L = 0.34\ \text{mm}$

From the relevant literature [5], it is known that the initial pressure drop of the backwash filter should be less than 0.015MPa, so the pressure drop of the fouling backwash filter is within the range of experience standard, and the filter can work normally.

(2) Calculation of pressure drop of filter cake

The filter cake formed on the surface of the filter screen of the fouling backwash filter is a compressible cake. The calculation of the pressure difference at both ends of the filter is a very complicated process. It involves the working conditions of the whole filter, including the processing quantity, the physical property of the suspension and the material properties of the filter cake formed [6]. Because of the compressibility of the filter cake, the final pressure difference between the two ends of the filter is more difficult to estimate [7] because of the change of its properties in the filtration process. The relevant literature [6, 7] is based on Darcy's law. Firstly, the filtration process of incompressible filter cake is studied. Then, starting from the microcosmic of compressible filter cake filtration, the formula for calculating pressure difference between the two ends of the compressible filter cake is obtained:

$$\Delta P = \frac{\mu K C V^2}{2A^2 t} \tag{5.33}$$

where:

- ΔP the total pressure drop (Pa) of fluid through cake;
- C response parameters;
- A the cake surface area perpendicular to the liquid flow (m^2);
- K through degree;
- V the volume of filtrate flowing within the filter cake in time t (m^3);
- μ Fluid viscosity (Pa. S);
- t Filter cake filtration time (s).

The transmission is as follows:

$$K = \frac{K''(1-\varepsilon_{av})^2 S_p^2}{\varepsilon_{av}^3} \tag{5.34}$$

where:

- K'' numerical constant;
- S_p The specific area of particles $\left(m^{-1}\right)$;
- ε_{av} The average porosity of the filter cake.

Usually $K'' = 5, \varepsilon_{av} = 0.807$, $S_p = \frac{S_{颗粒}}{V_{颗粒}} = \frac{4\pi r^2}{\frac{4}{3}\pi r^3} = \frac{3}{r} = 1.2\times10^5 m^{-1}$. So, Eq. (5.34) gives us: $K = 5.1 \times 10^9$

$$C = \frac{\rho_f c}{(1-\varepsilon_{av})\rho_s - \varepsilon \rho_f c} \tag{5.35}$$

where:

c percentage of solid content in suspension;
ρ_f The density of the suspension liquid;
ε porosity of filter cake;
ρ_s The density of the filter solid.

Take, $\rho_s = 2.5g/cm^3$, $\rho_f = 1.3g/cm^3$, so $c = 0.05\frac{\rho_s(1-\varepsilon_{av})}{(1-\varepsilon_{av})\rho_s+\varepsilon_{av}\rho_f} = 0.016$

The pressure drop ($\Delta P_{cake} = 0.032MPa$) at the two ends of the cake at unit time (1 h) can be obtainedvia Eqs. (5.35) and (5.33).

(3) **Total pressure drop of filter**

The total pressure drop is the sum of filter pressure drop and filter cake pressure drop. That is ($\Delta P_{total} = \Delta P_{net} + \Delta P_{cake} = 0.047MPa$). The pressure difference of the existing fouling backwash filter is generally 0.03–0.05 MPa, so the calculation in this case is in consistence with the experience standard. It can be seen that 0.047 MPa can be used as the initial pressure difference of differential pressure controller when seawater is filtered. When the filter is blocked, the pressure difference in the filter increases gradually. When the pressure difference reaches the initial set pressure difference more than 0.047 MPa, the differential pressure controller sends out the signal to start self-cleaning and when the pressure difference falls under 0.012 MPa, it indicates the completion of self-cleaning.

5.1.2 Design of a Swirl Filter

The design of hydrocyclone refers to the process of determining the structural parameters of each part of the hydrocyclone according to the relevant technical requirements such as processing capacity and separation efficiency. Firstly, according to the influence of various structural parameters of hydrocyclone on its processing capacity and separation efficiency, the basic dimensions of each structure are preliminarily determined. Finally, the optimum structural parameters are determined by numerical simulation. The hydrocyclone involved in this example is mainly used for the separation of water and sediment in the ballast water of the ship. Its basic structure is shown in Fig. 5.5 and its related design parameters arelisted in Table 5.4.

1. **Determination of main diameter**

In the design of hydrocyclone, we must first determine its main diameter (the diameter of the column section), and all other structural dimensions are determined by a certain proportion of the main diameter. The main diameter mainly affects the productivity and separation granularity of the hydrocyclone. Generally, the handling capacity and separation particle size increase with the increase of the main diameter. Large diameter swirler should be chosen when the design requirement is coarse and the processing capacity is high. On the contrary, small diameter should be chosen. According to the processing capacity of the design requirements, the main diameter

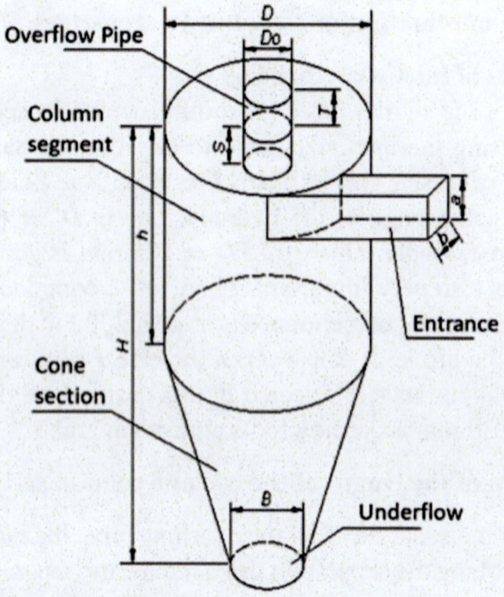

Fig. 5.5 The basic structure of hydrocyclone

Table 5.4 Preliminary design structure sizes of Hydrocyclone

Treatment volume $q_m/\text{m}^3\cdot\text{h}^{-1}$	Liquid density ρ_l/kgcm^{-3}	Sand density ρ_s/kgcm^{-3}	Sand volume fraction $\alpha_s/\%$
200	998.2	2650	1–10

of the hydrocyclone is calculated by using the formula of the main diameter of the hydrocyclone derived from Pang Xue Shi [8] (5.36).

$$D = \frac{1.95 q_m^{0.5} \rho_s^{0.25}}{\Delta p_m^{0.25} [C_w + \rho_s (1 - C_w)]^{0.25}} \tag{5.36}$$

where:

D the main diameter (cm);
Δp_m feed pressure (MPa), = 0.2 MPa;
q_m The amount of treatment (m^3/h);
ρ_s Sand grain density (kg/m^3);
C_w The concentration of sand particles (%), = 12.24%.

When the known parameters are substituted into the formula (5.36), the calculated diameter is 420.87 mm.As a result, $D = 420$ mm is selected.

2. Determination of other major structural parameters

(1) Determination of inlet structure and size

The inlet size of the swirler has a great influence on its separation performance. In general, increasing the inlet size helps to improve the handling and separation efficiency, but the excessive size will backfire. Pang Xue Shi [8] recommended that the best inlet diameter range of the hydrocyclone is $D_i = (0.15-0.25)\ D$. In the case of this design example, $D_i = 0.25D = 105$ mm is chosen. The shape of the entrance section is also very important. There are 2 common types of circular and rectangle, while the rectangular entrance is more helpful to the fluid expansion along the cyclone than the circle, it can weaken the effect of turbulence. Therefore, the rectangle tangential entrance is selected in this design, and the rectangular section size is 65 mm × 130 mm according to its equivalent area.

(2) Determination of the length of the cyclone column section

On the premise of the same depth of the overflow pipe, the increase of the length of the column can prolong the separation time, reduce the separation size, and increase the separation efficiency, but at the same time it will reduce the amount of treatment and increase the energy consumption. Because the column section of the cyclone is an effective centrifugal settlement area [9], therefore, it is suggested that the cyclone separator for solid liquid separation should take a larger column length, usually $H_c =$ (0.2−1.5), and this example is designed to take $H_c = 1.0\ D = 420$ mm.

(3) Determination of the diameter of the overflow

The size of overflow port has great influence on the handling capacity, separation efficiency, and separation particle size of hydrocyclone. Generally, with the increase of overflow diameter, the treatment capacity and separation particle size will also increase, but the rate of overflow will increase, resulting in the reduction of separation efficiency. In addition, the overflow pressure drop decreases with the increase of the overflow diameter. In order to reduce the energy consumption, the larger overflow diameter should be selected on the basis of ensuring the separation efficiency, usually (1.25−1.50) D_i [8]. This design takes $D_o = 1.35D_i = 141.75$ mm and was rounded to 140 mm.

(4) Determination of the diameter of the underflow opening

The diameter of the bottom port of the hydrocyclone has a significant effect on its separation performance. With the decrease of bottom outlet, the overflow particle size is thicker and the overflow rate increases, and the separation performance decreases. The size of the bottom port of the solid-liquid separation hydrocyclone is usually smaller than that of the overflow port, and the ratio should be reasonably chosen, generally $0.15-1D_o$. Because of the large volume of hydrocyclone designed in this case, in order to avoid blockage of the bottom outlet, the larger diameter underflow port should be selected, such as $D_u = 0.65D_o = 91$ mm, which was rounded to 90 mm.

Table 5.5 Preliminary design structure parameters of Hydrocyclone

Parameter	Size/mm	Parameter	Size/mm
The insertion depth	420	Overflow pipe of the main diameter of the cyclone	250
Inlet size	65 × 130	Bottom diameter	90
Overflow pipe diameter	140	Length of column	420
Overflow tube wall thickness	15	Cone angle	15°

(5) Dertermination of overflow tube insertion depth

When the overflow tube is too shallow, the processing capacity of the cyclone increases, the separation size becomes thicker and the separation efficiency is reduced. When the insertion is too deep, the small particles along the wall increase, which leads to the decrease of the separation efficiency of the small particles, but can improve the separation efficiency of the large particles. Pang Xue Shi [8] recommended the depth of insertion of the overflow pipe was $H_0 = (0.3-1)\ D$. In the example design, we chose $H_0 = 0.6D = 252$ mm, which was rounded to 250 mm.

(6) Determination of the wall thickness of the overflow pipe

Increasing the thickness of the overflow pipe can not only improve the separation efficiency of the cyclone, reduce the energy loss, but also improve the handling capacity of the cyclone [10]. Therefore, the thickness of the overflow pipe can be appropriately increased in design. The wall thickness is initially set to 15 mm in this design.

(7) Determination of the cone angle of the cyclone

For the same cyclone, the larger the cone angle, the shorter the cone and the shorter separation time for the solid phase. On the contrary, the smaller the cone angle, the longer the cone and the longer the separation time of the solid phase. The separation time will affect the processing capacity and separation particle size of the hydrocyclone, and the separation particle size will become coarser with the increase of cone angle. In general, the cone angle of the solid-liquid separation hydrocyclone ranges from 5° to 20°. In view of the fact that the cyclone designed in this example requires a fine separation particle size, the primary cone angle is 15°.

The structural size of the preliminary design of the hydrocyclone is shown in Table 5.5.

5.2 Design of Ultrasonic Cleaning Device

There are two kinds of effects of ultrasonic purification in the ballast water control system, one is to use the cavitation cleaning function of the ultrasonic air cavitation, to apply the mechanical shear force to the filamentous algae, which are lighter than water, or the hard algae spores in the water of the swirling filter, in order to use cavitation microbubbles to break the mycelium or spores. The destruction of the subshell creates conditions for the UV killing.The second is the cleaning ability of the oscillation shock produced by the ultrasonic wave, and the ultrasonic cleaning of the quartz protective sleeve of the ultraviolet lamp tube to prevent the deposit on the quartz casing caused by the heating of the ultraviolet lamp tube or the death microorganism on the quartz casing. The transmittance of the English bushing enables the ultraviolet radiation range to reach as far as possible.

5.2.1 Design of Ultrasonic Cavitation Reactor

1. **Ultrasonic transducer frequency determination and selection**

The frequency of ultrasound is determined according to the purpose of the reactor for water treatment. If an ultrasonic transducer is used to produce ozone or hydroxyl radicals, the transducer with a strong cavitation effect between 15–20 kHz should be selected. Through the cavitation, a series of pyrolysis reactions are produced in water, and active substances or hydroxyl radicals are generated. Ultrasonic cleaning accuracy of this frequency segment is poor, and even damages the object to be cleaned. If the mechanical effect or acoustic flow effect of ultrasonic transducer is used, the transducer with higher frequency should be used, such as 28 kHz. The transducer has less cavitation effect and does not produce living substances, and its cleaning accuracy is much higher than that of 20 kHz.

The size of the end surface of the ultrasonic transducer used in the ultrasonic cavitation purification reactor is directly related to the cavitation yield of ultrasonic cavitation. Larger area of ultrasonic radiation is more effective for increasing cavitation cavitation yield. Therefore, for the structure design of ultrasonic cavitation cavitation reactor, in order to improve the rate of cavitation cavitation, the larger area of ultrasonic irradiation is chosen as far as possible.That is, the surface area of the transducer selected under the same design power should be as large as possible. Although this will cause an ultrasonic transducer to radiate the average sound intensity of the sound wave in the reactor, the larger area of the ballast water can be irradiated by ultrasound, thus increasing the cavitation reaction yield.

The ultrasonic cavitation reactor is mainly used to do pre-treatment before UV sterilization. The microbes of larger bacteria micelles or macromolecular particles are split or cut and cleaned for their own tubes. Since cleaning is not the leading factor, the intermediate frequency ultrasonic transducer is generally selected to achieve the effect of ultraviolet disinfection before treatment. The ultrasonic transducer of type HNC-4SH-5028 was usually chosen, whose specific parameters are shown in Table 5.6.

Table 5.6 HNC-4SH-5028 transducer specific parameters

Model (Type)	Resonant frequency (kHz)	Static capacitance (PF)	Resonant resistance (Ω)	Shape size Diameter * height	Power (W)	Insulation impedance (2500VDC)
HNC-4SH-5028	28 ± 0.5	9600 ± 10%	≤20	67 mm × 68 mm	100	≥100 MΩ

2. **Determination of reactor structure type**

There are many types of reactor structures, such as trough type ultrasonic reactor, probe reactor, and cup reactor, but these reactors are all for dead water, so they are not suitable for the pre-treatment of the flow of ballast water. Figure 5.6 shows a reactor for dynamic treatment of flow water, but its ultrasonic probe is immersed in water, easily corroded by sea water, and its ultrasonic transducer has a short life. Therefore, this type is not suitable for the development of ballast water preprocessor. The reactor in Fig. 5.7 designed the inlet and middle cooling tube that added the treatment agent

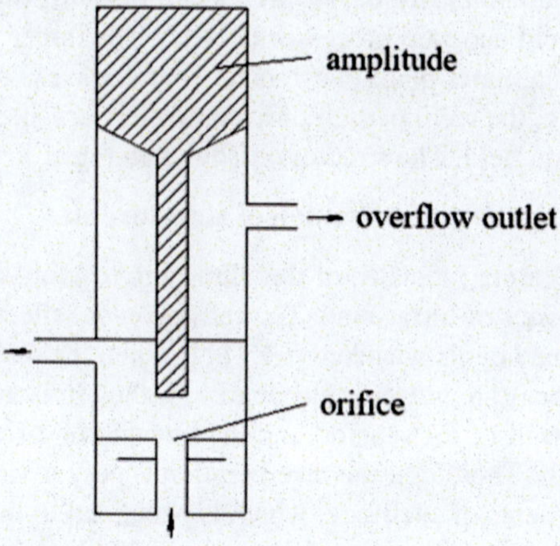

Fig. 5.6 Flow-type reactor

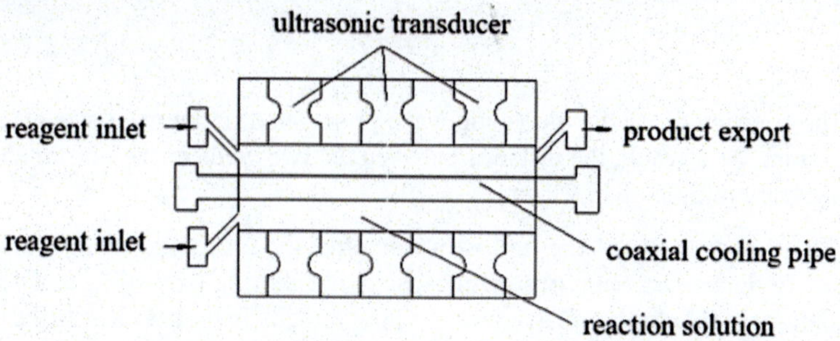

Fig. 5.7 Multi-probe-type ultrasonic reactor

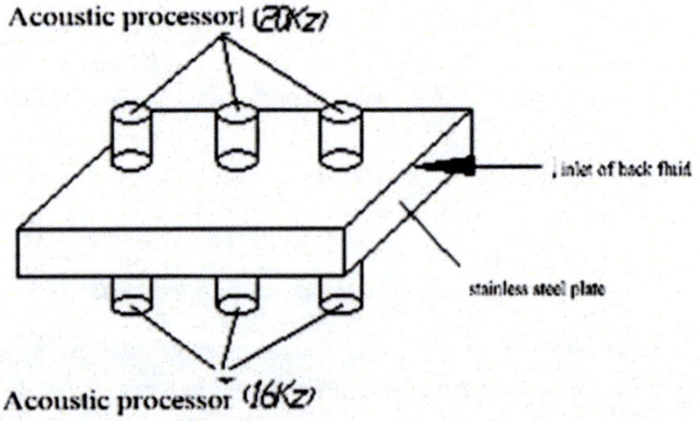

Fig. 5.8 Near-field Acoustic processor

into the flow water. Because the ballast water prevention system used pure physical treatment without adding any active material, the reactor was also not suitable.

Considering the particularity of the water in the ultrasonic cavitation reactor and the standard of the treatment effect, the near-field supersonic flow reactor is selected. It is a form of high intensity ultrasonic cavitation purification reactor, which is also called near field acoustic processor (Lewis). The upper and lower ends of the processor contain a metal plate that radiate sound waves. As the solution passes through the middle, the solid particles are completely crushed under the action of a strong reverberation field. The structure is shown in Fig. 5.8.

3. **Factors affecting the determination of structure size**

(1) **Wavelength factor**: : the size of the ultrasonic reactor is closely related to the various characteristics of ultrasound. The refraction or reflection of ultrasound in a reactor is determined by its wavelength λ [11]. When the distance between the two walls of the reactor is an integer multiple of λ/4, the ultrasonic wave will have full reflection on the wall of the reactor, and the attenuation of ultrasonic wave in the liquid is very small. Therefore, the reverberation is easy to occur in the reactor that meets the requirements of such size, which is beneficial to the occurrence of hyper acoustic cavitation. The frequency of the selected 28 kHz ultrasonic transducer and the propagation speed of sound wave in water 1480 m/s are substituted into the following formula (5.37):

$$c = \lambda f \tag{5.37}$$

The wavelength of the ultrasonic wave propagated in the reactor is obtained as 52.84 mm. As a result, the distance between the two walls of the reactor should be an integral multiple of 13.21 mm.

(2) **The flow velocity of the ballast water pipe**: the ultrasonic cavitation cleaning reactor of the ballast water treatment capacity is 300m^3/h. According to the relevant regulations of the ship's ballast water, the flow velocity in the seawater in the pipeline is not more than 3 m/s when the load is loaded or unloaded. Under this regulation, according to the standard size of pressure vessel in GBT9019 2001, it is known

that the path of the pipeline used in the 300 m^3/h ballast water treatment system is DN200, and its circulation cross section area is 37650 mm^2. According to the size of the pipeline, the cross section of the cylinder should not be less than 37650 mm^2.

(3) **The volume factor of ultrasonic irradiation**: In recent years, some scholars have divided the ultrasonic cavitation reactors into two typical structures according to their different structures: (1) continuous stirred tank reactor (CSTR); (2) piston flow reactor (PFR). The kinetic characteristics of the reaction solution can be deduced from the above two classical reactor theories. The CSTR model can be obtained as follows.

$$dC_R(t)/dt = \frac{Q}{V_S}[C_R(t) - C_S(t)] + \gamma_M r(K_{gc}) \tag{5.38}$$

where:

$C_R(t)$, $C_S(t)$	the instantaneous velocity of reactants and products, kg/m^3;
Q	reaction mixture volume, m^3;
V_S	volume of ultrasonic irradiation in ultrasonic air hole purification reactor, m^3;
γ_M	stoichiometric number;
$r(K_{gc})$	The reaction series.

Most of the ultrasonic cavitation purification reactions are first order reactions, and the non-integer reaction occurs only in the process of polymer degradation, and the probability of such reaction in this process is very small. After appropriate mathematical transformation is applied to Eq. (5.38), the formula of reaction rate constant in ultrasonic cavitation purification reactor under CSTR model is obtained as follows.

$$K_{obs} = \frac{V_S}{V_S + V_R} g K \tag{5.39}$$

where:

V_R reaction volume, m^3.

For the piston flow ultrasonic cavitation purification reactor, the material balance equation is obtained by using the correlation theory:

$$\left[\frac{\partial C_S(Vgt)}{\partial t}\right]_{V=const} = -Q\left[\frac{\partial C_S(Vgt)}{\partial t}\right]_{t=const} + \gamma_M r(Vgt) \tag{5.40}$$

After applying appropriate mathematical transformation to Eq. (5.40), the following formula can be derived:

$$C_R(t) = C_{D}g \exp\left[-\left(\frac{V_S}{V_S + V_R}\right)Kt\right] \tag{5.41}$$

According to formula (5.41), a mathematical formula for the reaction rate of ultrasonic cavitation purification in the PFR reactor can be obtained as follows.

$$K_{obs} = \frac{V_S}{V_S + V_R} K \tag{5.42}$$

According to the initial design conditions, the ultrasonic reactor is a piston flow ultrasonic cavitation purification reactor. Therefore, according to eqation (5.42), we can see that in order to improve the ultrasonic cavitation yield K_{obs}, it is necessary to improve the V_s, that is, to improve the volume of the liquid under supersonic radiation in the ultrasonic cavitation purification reactor.

(4) **The shape factors of the bottom wall of the reactor**: Usually, the ultrasonic cavitation yield is closely related to the bottom shape of the reactor, but so far there is not a very detailed description. According to the qualitative analysis of the relationship between the sound field reverberation mechanism and the electrical acoustic efficiency, we can see that most of the reactors are designed to be near field, which ensures that there is enough reverberation in the reactor, so as to ensure the normal occurrence of cavitation. In general, the acoustic energy density of incident wave on the surface of transducer *x cm* is:

$$\varepsilon = \begin{cases} \frac{I_0}{C} e^{-2ax} \\ 0 \quad \frac{x}{C} + T \le t < \frac{x}{C} \end{cases} \tag{5.43}$$

where:

C sound speed, m/s;
a attenuation coefficient of sound pressure, m^{-1}.

The sound energy density of the sound wave after N reflectionscan be calculated as follows:

$$\varepsilon_N = \frac{I_0}{C} e^{-2aNH}$$

Assuming that the rate of formation of cavitation bubble dn/dt is directly proportional to the K power of the acoustic energy, the relationship between the acoustic chemical yield D (T) and the sound energy can be deduced as follows:

$$dn/dt = A[\varepsilon(t)]^K \tag{5.44}$$

After making a proper transformation to Eq. (5.43), the following formula is derived:

$$D = BT \frac{A}{T_0} \int_0^{T_0} [\varepsilon(t)]^K dt \tag{5.45}$$

Fig. 5.9 Reactor configuration

According to the above theory, the bottom surface of the reactor should be designed as a rotating surface type, and as far as possible, no sharp corners and angular joints should be formed to prevent diffuse scattering of the sound field and to avoid the reduction of the local cavitation effect strength. It is also conducive to the realization of the processing technology.

Considering all the above factors, the cylinder body model is defined as a rectangular tube with a cross section as a square, and the size of the cross section is 264 mm × 264 mm, which is just 35 times that of the size $\lambda/4$, which provides favorable conditions for the homogenization of sound field. Considering the connection between the reactor and the whole ballast water system and the size required for the installation of the ultrasonic transducer array, the length is set at 630 mm. Because the ultrasonic reactor is mainly used to treat seawater, it is necessary to use stainless steel with strong chemical stability as its manufacturing material. Considering the efficiency of ultrasonic transmission and reverberation effect, the thickness of reactor wall should not be too large, generally only 2–3 mm. Therefore, the wall thickness of the reactor is set to 3 mm. The structure is shown in Fig. 5.9.

The cross-section area of the tube is larger than that of the pipeline, but when the ballast water passes, the water velocity is still very large, at 1.063 m/s. However, the cavitation effect or shear effect of ultrasonic wave requires certain dead water area or smaller flow velocity (generally less than 0.2 m/s), only in this way the ultrasonic can play its role. In order to promote the propagation of ultrasound in the water, a layer of vibration transmission network can be added in the reactor. The network is slightly larger than the size of the pipe. When the water passes through, the water can be fully mixed with the dead water in the reactor through the diversion of the net, so that the ultrasonic cavitation bubble can be entered into a larger flow velocity through the vibration transmission network. In the water body, the energy is released to achieve the treatment of larger microorganisms. The structure of the vibration transmission network is shown in Fig. 5.10.

4. **The design of the ultrasonic power of the reactor**

In order to improve the efficiency of electro acoustic efficiency, it is necessary to match the sound intensity I with the liquid volume. According to the design experience, when the liquid depth of the reactor is less than 400 mm, and the ultrasonic

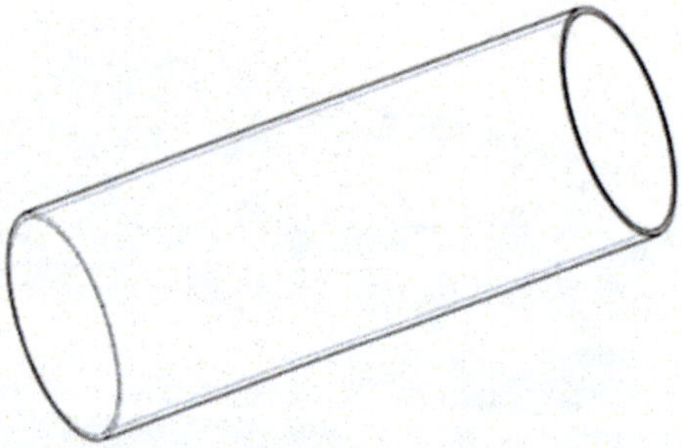

Fig. 5.10 3D model of preaching vibration grid

transducer is installed as a whole, the ultrasonic power should be used to designed by using the area power density parameter (sound intensity *I*) [12]. Within a certain limit, increasing the sound intensity *I* will increase the yield of ultrasonic cavitation purification reaction, but it cannot exceed that limit. Generally speaking, when the sound intensity *I* is greater than 3 W/cm^2, the yield of ultrasonic hole will not increase significantly. Taking into account factors such as cost and life span, sound intensity should be controlled within a certain range. For example, the increase of acoustic intensity in the transition will increase the power consumption of the ultrasonic transducer, and at the same time, it will also have great adverse effects on the life of the ultrasonic cavitation purification reactor. On the contrary, if the ultrasonic transducer with a small end surface is used, the sound intensity of the seawater is relatively large, but the rest of the seawater cannot be irradiated by ultrasonic, which cannot improve the rate of ultrasonic cavitation. Therefore, in order to improve the rate of ultrasonic cavitation and reduce the power consumption of ultrasonic transducer, the surface area of ultrasonic transducer used by ultrasonic cavitation cleaning reactor should be as large as possible, making the mean sound intensity of *I* less than $I \leq 3\ W/cm^2$.

As the cross section of the reactor is square, there are four equalreactor walls. Therefore, the area of the reactor can be obtained by calculating the areas of two of the surfaces. According to the specific size of the reactor, it can be obtained as follows.

$$A = 2 \times (C - 2R)L = 3099.6\ \mathrm{cm}^2$$

where:

C the length of the cross section of the reactor, C = 286 mm;
R the radius of the corner of the reactor, R = 20 mm;
L the length of the reactor, L = 630 mm.

In the pretreatment process of ballast water, the main effect is to cut off the larger microbes and bacteria micelles, so the required sound strength is less than the sound intensity used to degrade organic matter. According to the experimental experience, the sound intensity used to treat microorganisms should be around 2 W/cm^2. In this design, the sound intensity is set at $I = 2.3$ 2 W/cm^2. Therefore, the total power of the ultrasonic transducer is as follows.

$$P = A \times I = 3099.6 \times 2.3 = 7129.08 \text{ W}$$

According to the calculation of total power, it is rounded to 7200 W, taking into account the loss of energy. According to the parameters of the selected transducers, 72 ultrasonic transducers are required to meet the power requirements.

5. **Design of the structure of ultrasonic cavitation reactor**

(1) **Installation of ultrasonic transducer**

The installation form of the ultrasonic transducer is generally divided into three types: direct mounting type, box mounting type, and vibration plate type [12].

The direct installation means that the ultrasonic transducer is directly fixed on the outer wall of the reactor. This type of installation is simple in structure, is convenient for maintenance and replacement, and is suitable for small and medium sized ultrasonic reactors.

The vibration plate type means that the ultrasonic transducer is distributed on a stainless steel plate according to the required array, and the stainless steel plate is used as the bottom of the reactor. This installation method combines the advantages of the above two ways, but the requirements for sealing are very high.

According to the design requirements of the ballast water treatment system, the design of the reactor structure can only choose the first type of installation, that is, the direct installation, as shown in Fig. 5.11.

There are two ways to connect directly installed transducer and container wall: flange connection and integral bonding.

The flange connection is first used to glue the ultrasonic transducer to a circular stainless steel plate (vibrating plate) with a special adhesive, and then open the hole at the bottom of the reactor. The diameter is less than the diameter of the vibrating plate. The ultrasonic transducer with a vibrating plate is placed on the bottom wall of the reactor, and the vibration plate is extended to the external of the reactor by

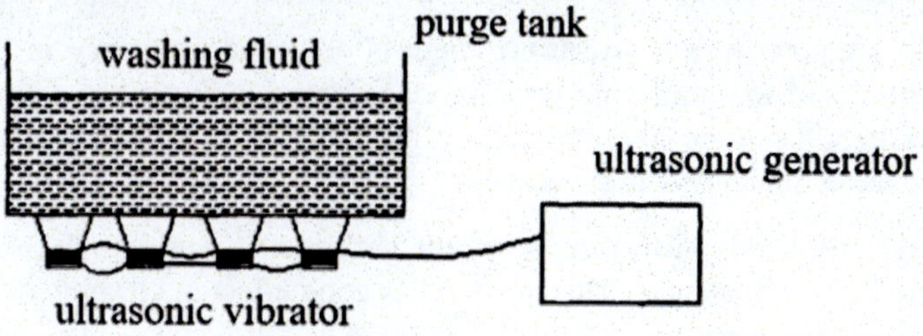

Fig. 5.11 The structure of transducer installed directly

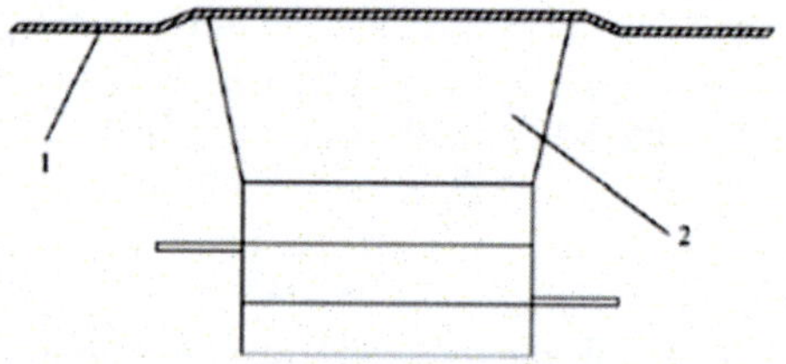

Fig. 5.12 Adhesive type structure as a whole

opening the hole. The sealing pad is inserted into the flange between the bottom wall of the reactor and then fixed by bolts, as shown in Fig. 3.21.

The overall adhesive structure is much simpler than the flange connection. In the outer wall of the reactor, the groove is pressed into the groove, the size of which cannot be too deep, usually controlled at about 5 mm, and the ultrasonic transducer is directly attached to the groove, as shown in Fig. 5.12.

Wang Haiou etc. have done a sound field test on these two forms of connection and found that the sound energy in the liquid of the overall adhesive reactor is obviously higher than that of the flange connection type [13]. For the flange connection structure, a portion of the energy loaded on the vibrating plate is passed directly to the reaction fluid, and the rest passes through the connecting fixings to the bottom wall and then passes through the bottom wall to the cleaning fluid. However, a nonrigid connection is formed between the vibration plate and the bottom wall, which forms a nonrigid connection between the vibration plate and the bottom wall, so that this part of the energy will be lost directly when the vibratory plate passes the energy to the liquid. For the whole adhesive type, the ultrasonic transducer is directly fixed to the bottom wall of the reactor. The vibration energy of the transducer is transferred directly to the bottom wall of the reactor, and then the bottom wall of the reactor passes the energy to the reaction liquid. There is no damping connection in the process, so there is no energy loss due to damping in the propagation of ultrasonic energy. As a result, the energy conversion efficiency of the latter is higher than that of the former.

(2) **Selection of ultrasonic generator**

According to the different vibration modes of ultrasonic generator, the ultrasonic generator can be divided into the following categories:

① **Single frequency rate vibration**: there is only one frequency in the process of vibration, and the ultrasonic wave produced by this mode has a strong effect. However, because of the single frequency, the sound waves converge, there may be some radiation blind areas in the reactor.

② **Change amplitude vibration**: when the generator is working, the amplitude of ultrasonic wave will change. This method has good effect in cleaning or degassing, but it makes a lot of noise.

③ **Interactive frequency change oscillation**: this mode can intermittently emit different frequencies of ultrasonic waves to drive ultrasonic transducers with different frequencies. At first, it was mainly to solve the problem of the blind area of the sound field. But experiments show that when the ultrasonic transducer with different frequencies is mounted on the same vibrating plate, the transducer with higher frequency is prone to failure. Because of the intermittence of the mode of vibration, the transducer only works half of the time, making the sound intensity weak and cannot achieve good treatment effect.

④ **Simultaneously multi-frequency oscillation**: The difference between the vibration mode and the third method is that multiple frequencies are simultaneously emitted. This multi-frequency ultrasonic wave produces a relatively uniform sound field while operating, but the sound intensity is very low.

⑤**Shaped energy oscillation**: It can be seen from the definition that the vibration mode is to multiply the emission portion of the vibration head through a tool head in a very small area, thereby multiplying the power density. This method was successfully applied to spinneret cleaning.

According to the over current characteristics of the ballast water resistant biological treatment device, the ultrasonic transducer with the same frequency is selected. In order to improve the processing effect of the ultrasonic reactor, it must have strong sound intensity, so the single frequency ultrasonic transducer is more suitable. Eliminating the blind area of acoustic field can be realized by the design of sound field.

(3) **Homogenization of sound field**

According to the working characteristics of the ultrasonic cavitation reactor, the ultrasonic transducer installed on the reactor can be regarded as the sound field versus the radiation pattern. Along the direction of the flow, the ultrasonic waves propagate on the two main reaction walls and form reflection on their respective surfaces. In the process, two rows of sound waves with the same frequency and the opposite direction of propagation are formed in the reactor. By choosing the origin of the common coordinates and the starting point of time, we can get the expression of the two series of sound waves as follows.

$$y_1 = A\cos(\omega t - kx)$$

$$y_2 = A\cos(\omega t - kx)$$

The combined displacement amplitude at the position where the two columns of sound wave meet is:

$$y = y_1 + y_2 = A\cos(\omega t - kx) + A\cos(\omega t - kx) = (2A\cos kx)\cos\omega t \tag{5.46}$$

Because $k = 2\pi/\lambda$, λ is the wavelength of sound wave propagating in water, k is substuituted into Eq. (5.46) to get the following:

$$y = \left(2A\cos\frac{2\pi}{\lambda}x\right)\cos\omega t \tag{5.47}$$

In the formula (5.47), $\cos\omega t$ shows that the superimposed sound wave in the reaction liquid is still changed according to the sine law, and $2A\cos\frac{2\pi}{\lambda}x$, the amplitude of each point, changes periodically with x.

When: $x = 0, \pm\lambda/2, \pm 3\lambda/2$

$$\text{That is}: \quad x = \pm i\lambda/2 \quad i = 0, 1, 2\ldots \tag{5.48}$$

Substitute it into Eq. (5.47), we get $\left|2A\cos\frac{2\pi}{\lambda}x\right| = 2A$, which shows that when the sound wave is propagated to the above position, the amplitude of the superposition is 2A, which is the maximum amplitude.That is, the position of the wave peak in the standing wave field [14].

When: $x = \pm\lambda/4, \pm 3\lambda/4, \pm 5\lambda/4, \ldots\ldots$

$$\text{That is}: \quad x = \pm(2i + 1)\lambda/4 \quad i = 0, 1, 2\ldots \tag{5.49}$$

It can be obtained via Eq. (5.47) that $2A\cos\frac{2\pi}{\lambda}x = 0$, which shows that when the sound wave is propagated to the above position, the amplitude of the superposition is 0, the trough position of standing wave field. According to formula (5.48) and (5.49), the distance between two adjacent crest peaks or two wave valleys is $\lambda/2$, while the distance between adjacent wave valleys and crest peaks is $\lambda/4$.

Therefore, the ultrasonic transducers are divided into two groups, and the arrays on the left and right sides are distributed on the outer wall of the reactor, thus forming a single direction, single frequency pairing. In order to achieve the mutual compensation of the amplitude of the ultrasonic waves, the standing wave peak formed by each of the transducers on one side should meet the standing wave trough formed by each transducer on the other side. Because the distance between the peaks of the standing wave and the trough is λ/4, the range of the transducer should be an integral multiple $\lambda/4$, in order to achieve an effective cleaning purpose. As shown in Fig. 5.13, the ultrasonic transducer is simulating the distribution of the sound field. Figure 5.14 shows the simulation of the differential distributed sound field.

It can be seen from the above figures that, regardless of the locations of ultrasonic transducers, as long as the distance between the end faces of the ultrasonic transducer is an integral of $\lambda/4$, the standing wave valleys of the transducers on both sides will be remedied by the standing wave peak on the opposite side, so the sound field in the reactor will become homogenized. The blind area of sound field caused by standing wave phenomenon will also be improved. The specific arrangement and internal structure of ultrasonic transducer and reactor are shown in Fig. 5.15.

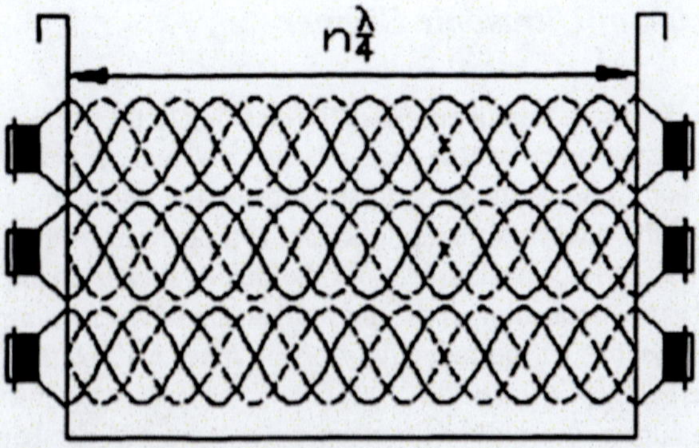

Fig. 5.13 Simulation of distributed sound field

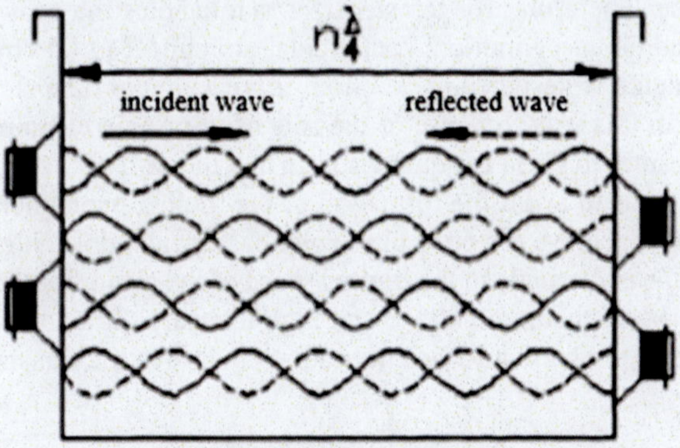

Fig. 5.14 Simulation of sound field with differential open distribution

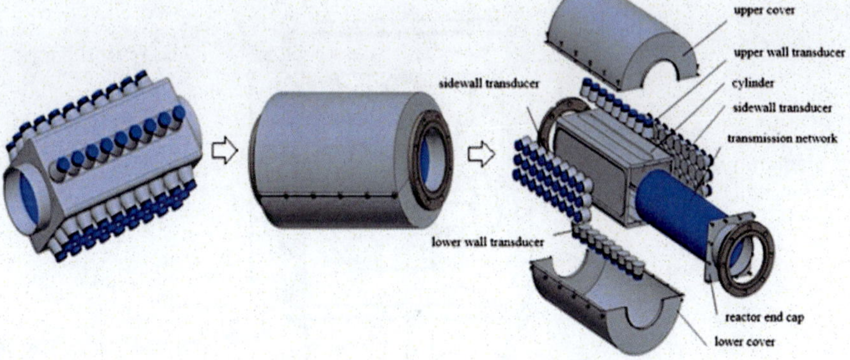

Fig. 5.15 US transducer arrangement and the reactor internal structure

5.2.2 Design of Ultrasonic Cleaner

The ultrasonic cleaner is mainly used for the cleaning of the ultraviolet bactericide in the marine ballast water control system. It is also a typical over current cleaning type. The purpose is to remove the mineral structure in the seawater caused by the high temperature of the ultraviolet lamp tube on the inner wall of the medium pressure high-strength ultraviolet bactericide and the quartz sheath of the ultraviolet lamp tube. As well as the deposition of the microorganism on it, ensuring the permeability of the quartz sheath and the reflectivity of the inner wall of the cylinder, the UV sterilization efficiency is the best.

Using the supersonic ultrasonic cleaner, the wire rod scraper mechanism for cleaning the ultraviolet lamp tube can be saved, which makes the structure of the germicidal device more concise besides the medium pressure high strength ultraviolet lamp tube. Therefore, for the sterilizer that handles the heavy flow of ballast water, the traditional cylindrical bactericidal structure can be changed in design, and the germicide is designed as a bottom-in, rectangular tube sterilizer, as shown in Fig. 5.16. In this way, because of the role of ultrasonic cleaning, the ultrasonic cleaning continues to clean the quartz sheath and the wall of the tube when the sterilizer works, so as to avoid the formation of the fouling phenomenon. At the same time, in order to achieve effective ultraviolet bactericidal dose, it can also lengthen the cavity of the bactericide in the limited space of the cabin, for the arrangement of the sterilizer. More medium voltage and high intensity ultraviolet tubes will allow the water stream to obtain a sufficient irradiation dose in the transient overcurrent.

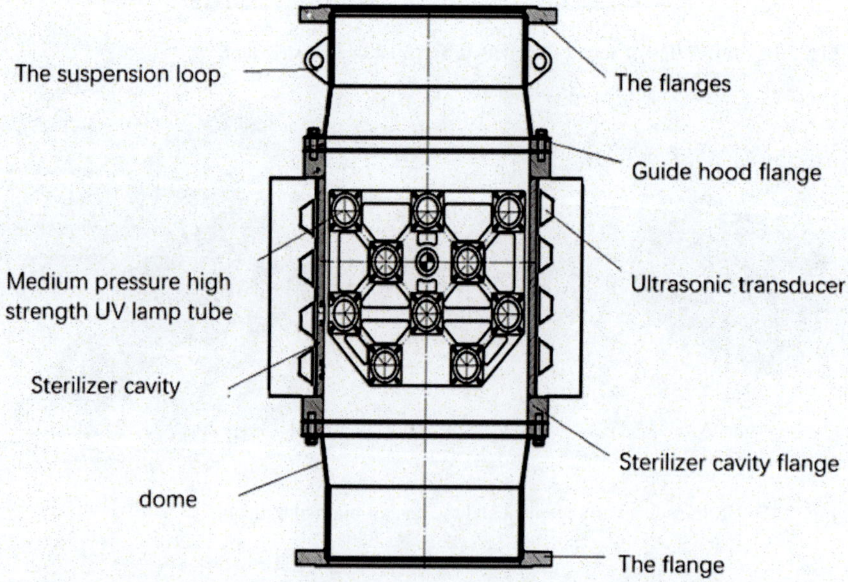

Fig. 5.16 With over-current rectangular tube type of ultrasonic cleaning sterilizer

The selection and design method of ultrasonic transducer for supersonic ultrasonic cleaner is similar to the design method of ultrasonic cavitation reactor. It is mainly that the ultrasonic transducer is arranged on both sides of the bactericidal device perpendicular to the ultraviolet lamp. The design calculation method is the same as the design calculation method of the ultrasonic cavitation reactor.

5.3 Design of Bactericidal Unit

Bactericidal unit is the key component in the marine bio management system of ballast water. The strength of its bactericidal ability is directly related to the product certification evaluation of IMO and USCG by the biological management system of ballast water, and the qualification of market access is obtained. Therefore, whether it is a bactericidal device that is evaluated by IMO G8 using the principle of pure physical method, or by using the principle of chemical additive method to accept the bactericidal device evaluated by IMO G9, the standard of its quality evaluation is the efficacy and effect of sterilization.

The principle of pure physical method refers to the direct use of ultraviolet, heating, mechanical, inert gas and other physical methods in the over flow treatment of ballast water, without any chemical reagents, such as the system of UVC and inert gas sterilization unit.The chemical additive rule refers to a system that uses electrolysis or photocatalysis to generate other chemicals or directly add chemicals to the treated seawater during the over-current treatment of ballast water, such as a system of optional photocatalytic, electrolysis or chemical sterilization unit.

5.3.1 Design of UVC Bactericidal Unit

UVC bactericidal mechanism mainly is to make the nucleic acid mutation, and then prevent the reproduction and transcription of a series of reproduction activities, directly damage the cells and inhibit the activity of the cells; the free radicals are produced by irradiation in the water, and the bacteria are killed directly.

When the sterilizer works, its interior is filled with water, and the UVC lamp tube cannot be directly contacted with the treated water. A layer of quartz sleeve is set on the outside of the lamp to ensure its safe use. When the UVC tube is working normally, its surface temperature is very high, such as medium pressure UVC tube, the normal working temperature is as high as 900°C. Therefore, quartz sleeves must be high temperature resistant and high transmittance. When UVC crosses the quartz tube into the water, the radiation intensity decreases with the increase of the irradiation radius of the light source. Its attenuation satisfies the exponential decay law, namely Lambert Law:

$$I = I_0 e^{-\alpha_w d} \tag{5.50}$$

Table 5.7 Determination of different irradiation radius of UVC in distilled water

Water depth (cm)	0.8	4.5	10
Penetration rate (%)	83	35	10

where:

I_0 initial irradiation intensity (W/cm^2);
α_w The absorption coefficient of light (cm^{-1});
d The thickness of water layer (cm).

The penetration ability of UVC decreases with the increase of irradiation radius in water. The UVC penetration rate in distilled water is shown in Table 5.7. The absorption coefficient of distilled water to UVC can be calculated by substituting the numerical valuesin the table into Lambert's law to get $\alpha_w = 0.2326\,\text{cm}^{-1}$. The calculation of subsequent UVC strength can be done with a reference to this value.

1. **Bactericidal kinetics of UVC**

The essence of UVC sterilization is photochemical reaction. However, when UVC is sterilized, the microorganisms will also have a series of photoreactivation or dark resurrection reactions, which in turn are biochemical reactions. Photochemical reactions dominated the two.

The rate equation of UVC bactericidal reaction is as follows.

$$r_A = -(dN_A/dt) = kN_A^n$$

On both sides, the logarithm can be obtained:

$$\lg[-(dN_A/dt)] = \lg k + n \lg N_A$$

where:

n reaction series;
k reaction rate constant.

UVC sterilization is generally first-order reaction, so n = 1. From this a kinetic model of UVC sterilization can be derived to meet the first-order reaction.

$$N = N_0 e^{-kIt} = N_0 e^{-kD} \tag{5.51}$$

where:

N the concentration of bacteria after ultraviolet irradiation (A/100 ml);
N_0 Initial body concentration (A/100 ml);
k The rate constant ($cm^2/W\cdot S$)), the recommended value of the EPA Design Manual of the United States is 0.067–0.38;
D Ultraviolet radiation dose (J/cm^2);

Table 5.8 The influence of chromaticity of UVC absorption coefficient

Chromaticity (degree)	1	5	10	15	20	25	30	35	40	45
Absorption coefficient (cm^{-1})	0.025	0.08	0.12	0.16	0.20	0.23	0.25	0.32	0.35	0.40

I radiation intensity of ultraviolet light (W/cm^2);
t irradiation time (s).

It can be concluded that the killing effect of UVC on microbes is closely related to the initial concentration of microorganism, and it is exponentially related to the irradiation dose of ultraviolet lamp. The corresponding mathematical transformation can be applied continuously to get the following:

$$\lg \frac{N}{N_0} = -kIt = -kD \tag{5.52}$$

After a large number of experiments, the applicability of the first order reaction formula to the virus killing rate can reach 99.999% (also known as 5 *lg*), and the applicability of the bactericidal rate of the bacteria can also reach 3 lg.

2. **Factors affecting the sterilization of UVC**

(1) **Water Quality**

Water quality has a great influence on the germicidal efficacy of UVC, which is closely related to the color, turbidity, and iron content of water. For chromaticity, the influence of chroma on UVC absorption coefficient is obtained through experiments, as shown in Table 5.8.

It can be seen from the above table that with the increase of chromaticity, the absorption coefficient of water to UVC is also increasing. According to Lambert's law, the attenuation of UVC is faster and faster with the increase of chromaticity.

The effect of turbidity on the germicidal efficacy of UVC is mainly manifested in that some turbidity substances in the water absorb, scatter or reflect UVC rays, thus affecting their penetrating power. Therefore, for some small microbes, they can be hidden in these turbidity material particles, avoiding the direct exposure of UVC and greatly reducing the bactericidal effect of UVC.

The germicidal efficacy of UVC is closely related to the iron content in seawater. The free iron has a strong absorption to UVC, so a part of the energy irradiated by UVC will be lost on the iron ions, causing a significant decrease in the dose on the bacteria. Therefore, it is necessary to consider the pretreatment of iron ions in seawater to ensure the bactericidal effect of UVC.

(2) **Temperature**

Temperature has a great influence on the UVC lamp. Because the product is developed by using medium voltage lamp tube, the normal working temperature of medium voltage lamp tube is far higher than that of low voltage lamp tube. According to

Table 5.9 G67T5VH lamp illumination under the condition of different voltage

Voltage (V)	Illumination range(μW/cm^2)	Average illumination(μW/cm^2)
200	110–117	142
210	135–180	165
220	160–210	185
230	165–225	202

the measurement, the surface temperature of the medium pressure lamp tube has reached 900 degrees Celsius. If the surrounding environment can provide such a high temperature, so that the UV lamp does not have any heat exchange at work, and its UVC conversion efficiency reaches 15.67%. But such conditions are impossible to achieve in the process of ballast water. Due to the existence of quartz casing, when the UV lamp is working normally, the air between the lamp surface and the quartz sleeve can be heated to about 200 degrees Celsius, the rest of the heat will be taken away by water, and in this situation the conversion efficiency of UVC is only 12.33%. Therefore, when calculating the dose of UVC, the effect of temperature must be taken into account.

(3) **Tube voltage**

The working voltage of the UVC lamp will directly affect its illumination. Therefore, only by ensuring the stability of the working voltage can the stability of the bactericidal effect be maintained. Through the voltage change experiment of the G67T5VH low voltage lamp, the data shown in Table 5.9 can be obtained.

It can be seen from the above table that the radiation intensity of UVC lamp is greatly influenced by voltage. Because the rated working voltage of the lamp is 230 V, and and for every 10 V drop, the illumination will be reduced by 12–20 μW/cm^2.

(4) **Other influencing factors**

In addition to the influencing factors described above, the UVC dose, the type and concentration of microorganisms, the time of use of the lamp, and the like can all affect the bactericidal effect of UVC.

3. **UVC sterilizer design calculation**

The main structure of UVC sterilizer includes: sterilizer cylinder, UVC lamp tube and casing, cleaning mechanism, and power structure. In the treatment of ballast water, the flow of water can be divided into three stages: influent, bactericidal, and effluent. According to the characteristics of the treatment process, the UVC sterilization should be defined as over-current.

The function of UVC sterilizer is to inactivate the smaller bacteria, viruses and algae in the ballast water, so it is an important component to ensure good treatment effect and is the core of the ballast water treatment system. Considering the limited space of the ship, the treatment effect of the sterilizer is required to meet the IMO ballast water treatment standard, and at the same time, efforts should be made to reduce the treatment time and make the sterilizer as small, practical, and efficient as possible.

(1) UVC lamp selection

A 300 m^3/h ship ballast water efficient treatment system was designed for land-based and sea-based experiments.

Select DQ3030 medium-pressure high-strength UVC mercury lamp, According to the size of the tube, the size of the quartz sleeve can be determined as 29 mm in inner diameter, 32 mm in outer diameter, and 430 mm in length.

(2) UVC irradiation dose

The irradiation dose of UVC is represented by D and refers to the product of the irradiation intensity *I* at a certain point and the irradiation time t. It is a parameter to measure the bactericidal capacity of UVC sterilizer. Its specific expression is:

$$D = I \times t \tag{5.53}$$

where:

I irradiation intensity to a certain point, μW/cm^2;
t irradiation time to a point, s.

For the entire UVC sterilizer, the radiation dose at different locations is different. Therefore, the sterilizer dose should be expressed as the average dose, i.e.:

$$D = I_{avg} \times \bar{t} \tag{5.54}$$

where:

I_{avg} the average UV irradiation intensity in the sterilizer, μW/cm^2;
$\bar{t}$ The average time required for the fluid to pass through the sterilizer, s.

In general, the average time of over-flow is determined by the effective volume and flow rate of the sterilizer. That is, it is expressed by the over-current distance and flow rate.

The radiation dose is the main parameter to measure the bactericidal capacity of the sterilizer. Therefore, the core issue of UVC sterilizer design is to determine the UVC irradiation dose. From formula (5.54), it can be seen that the dose size of UVC irradiation is related to the output power of UVC lamp, irradiation distance and irradiation time (Table 5.10).

(3) **UVC lamp arrangement**

There are two basic methods for UVC intensity calculation: Radial model and Point Source Summation (PSS) model.

Table 5.10 UVC dose needed to kill different microorganisms

Microorganism	Dose ($\mu J/cm^2$)	Microorganism	Dose ($\mu J/cm^2$)
E.coli	6000	Flu virus	6600
Typhoid bacillus	7600	Tetanus virus	22000
Bacillus spores	40000	Hemolytic Streptococcus	5500
Staphylococcus aureus	12000	Green Streptococcus	3800
Diphtheria	20000	E.coli phage	6600
Tuberculosis	20000	Salmonella	10000
Aspergillus flavus	9900	Shigella	4200
Aspergillus spores	600000	Vibrio cholerae	6500

The Radial Radiation Model:

The UVC lamp is considered as a linear light source with uniform energy. In the calculation, the following is assumed.

① The energy radiated by the UVC lamp is evenly distributed along the axial direction of the lamp tube;
② UVC rays are perpendicular to the axis of the lamp tube and the outer wall of the quartz tube, and radiate energy in the form of a cylindrical surface;
③ Only consider the absorption of UVC by contaminants in water, not the refraction of air, quartz and water interfaces, and the scattering and reflection of contaminants in water.

Point Source Summation Model:

The UVC lamp is considered to consist of a myriad of independent point light sources, and radiates outward uniformly in the form of a spherical surface. In the calculation, the following assumption is made:

① UVC lamp is composed of numerous isotropic point light sources;
② Each point light source is independent of each other and radiates energy outward in the form of a spherical surface;
③ Only the absorption of UVC by contaminants in water is considered, and the refraction of air, quartz, and water interfaces, and the scattering and reflection of contaminants in water are not counted.

According to the characteristics and requirements of the target product, the radial radiation model is more suitable for the design calculation of the device.

Since the cylinder of the designed sterilizer has a cylindrical shape and the lamp tube is arranged along the axial direction of the cylinder body, the coordinates are set in such a form that the axis of the cylinder body is the x-axis, and the center of the effective part of the sterilizer is the origin. And the plane perpendicular to the x-axis is the yoz plane, and two radii selected from the origin and perpendicular to each other are defined as the y-axis and the z-axis, respectively.

If there is only one UVC lamp in the sterilizer, its axis position coincides with the axis of the sterilizer. Based on the assumption of radial radiation, UVC radiation intensity at any point on the cylindrical surface of the sterilizer can be obtained as follows.

$$I_0 = \frac{P\eta}{2\pi r L} \tag{5.55}$$

where:

P UVC lamp input power, W;
η UVC lamp conversion efficiency to UVC;
L Effective length of UVC lamp, m;
r Any cylinder radius along the lamp axis, m.

Since the UVC lamp cannot be in direct contact with water during normal operation, it needs the protection of the quartz sleeve. As a result, when the UVC passes through the quartz sleeve, there will be a part of the power loss. The loss size is related to the absorption rate of the quartz sleeve. After passing through a quartz tube, the UVC enters the water, and the law of propagation in the water satisfies the previously described Lambert's law. Therefore, the mathematical expression of UVC propagation in water can be obtained as follows.

$$I = I_0\eta_q e^{-\alpha_w d_w} = \frac{P\eta\eta_q}{2\pi r L}e^{-\alpha_w d_w} \tag{5.56}$$

where:

η_q UVC penetration of quartz sleeves;
d_w The thickness of the water layer, m.

$d_w = r - R_q$, r is an arbitrary cylinder radius along the axis of the lamp tube and is the outer radius of the quartz tube.

According to formula (5.56), $I_{\min}$ will appear on the cylinder surface furthest from the axis of the lamp tube. For a single tube, the specific location of $I_{\min}$ is the cylindrical surface of the inner wall of the reactor.

For its average irradiation intensity $\vec{I}$, from the assumptions of the radial radiation model, it is known that the average radiation intensity in any plane along the x-axis is equal. Therefore, the average irradiance of the entire sterilizer is equal to the average irradiance at any x-section, which is $\overline{I}(x)$

Since the UVC sterilizer has an arbitrarily symmetrical axial cross section, it is only necessary to calculate the 1/8 circle when finding the average cross-section light intensity. Convert rectangular coordinates to cylindrical coordinates, i.e., $x = x$, $y = r\cos\theta$, $z = r\sin\theta$. Since the irradiating power is equal to the product of light intensity and surface, the following can be obtained:

$$I(x, r, \theta) \cdot dA = \frac{p\eta\eta_q}{2\pi r L}e^{-\alpha(r-R_q)}rdrd\theta \tag{5.57}$$

Integrating both sides of Eq. (5.57), we have the following:

$$\bar{I}(x)\cdot A=\int\limits_{\theta_1}^{\theta_2}d\theta\int\limits_{r_1}^{r_2}\frac{p\eta\eta_q}{2\pi rL}e^{-\alpha(r-R_q)}rdr \tag{5.58}$$

According to the above boundary conditions, $A=\frac{\pi}{8}(R^2-R_q^2)$, $\theta\in\left(0,\frac{\pi}{4}\right)$, $r\in(R_q,R)$, R isthe radius of the inner wall of the reactor. Subtituting them into formula (5.58), the following is obtained:

$$\begin{aligned}\bar{I}(x)&=\frac{8}{\pi(R^2-R_q^2)}\int\limits_0^{\frac{\pi}{4}}d\theta\int\limits_{R_q}^{R}\frac{p\eta\eta_q}{2\pi rL}e^{-\alpha(r-R_q)}rdr\\&=\frac{p\eta\eta_q}{\pi(R^2-R_q^2)L}\cdot\frac{1}{0.2326}[1-e^{-\alpha(R-R_q)}]\end{aligned} \tag{5.59}$$

Formula (5.59) is the average irradiation intensity of a single UVC lamp sterilizer.

In order to ensure the bactericidal effect of the sterilizer, the average sterilizing dose of the sterilizer designed in this case should be more than 250,000 μJ/cm^2. Taking into account the high efficiency of sterilization, 14 medium-pressure UVC mercury lamps DQ3030 were selected and arranged in two cylindrical surfaces. Continue to use the form established by the above coordinates, the specific arrangement is shown in Fig. 5.17.

For the calculation of the illuminance of the lamps in the multiple sterilizers, the illuminance at any point should be the cumulative value of the illuminance of each lamp at this point. However, when the number of lamps is large, some occlusion will

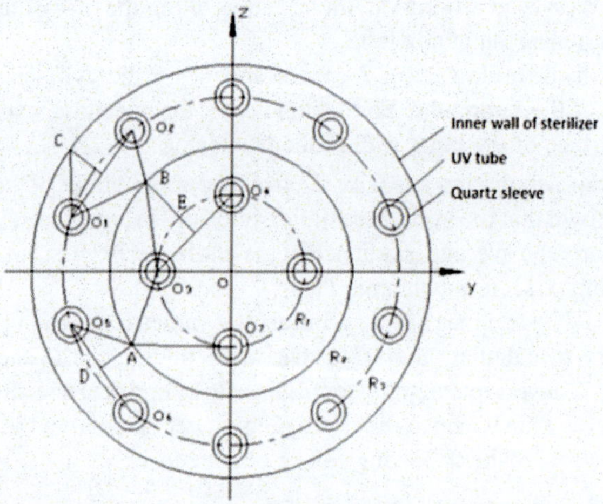

Fig. 5.17 UVC lamp profile

inevitably occur. Since the UVC irradiation intensity satisfies Lambert's law, when the distance exceeds a certain value, the contribution to the illuminance at this point is negligible. Therefore, the lamp layout for the target sterilizer can be specified as follows.

(1) In a region with a radius less than R_1, the UVC radiation intensity at any point is the cumulative irradiation intensity of the four lamps of the array on the circumference of the inner circle. The point with the weakest light intensity is the O point.
(2) In the area between R_1 and R_3, the UVC radiation intensity at any point is the cumulative intensity of the four lamps near that point. Theoretically, the weakest point of light intensity should be on the center of a quadrilateral composed of any four tubes. To facilitate calculation, points A and B are used as reference points for minimum illumination intensity.
(3) In the area between radius R_3 and the inner wall of the sterilizer, the UVC irradiation intensity at any point is the cumulative intensity of the two lamps near that point. The point with the weakest light intensity is the point C. According to the conditions specified above, determine the circumferential dimensions of the two lamps.

① Design of the inner radius of the lamp tube

Because the four lamps are symmetric about the origin in the coordinate system, the polar coordinates represent the coordinates of the four lamps. The coordinates of the axis of the i-th lamp among the four lamps are $\left(R_1\cos\left(\frac{i-1}{4}2\pi\right), R_1\sin\left(\frac{i-1}{4}2\pi\right)\right)$, The ultraviolet intensity at any point in the circle with radius R_1 can be obtained with formula (5.56):

$$I_i(x, y, z) = \frac{P\eta\eta_q}{2\pi r L}e^{-\alpha_w(r_i - R_0)} \tag{5.60}$$

The total UV intensity of any of the 4 lamps in the area is:

$$I_1(x, y, z) = \sum_{i=1}^{4}\frac{P\eta\eta_q}{2\pi r L}e^{-\alpha_w(r_i - R_0)} \tag{5.61}$$

where r_i is the distance from the point (x, y, z) to the axis of the UV lamp, $i =$ 1,2,…4. In the interface of x = C:

$$r_i = \sqrt{\left[y - R_1\cos\left(\frac{i-1}{4}\right)2\pi\right]^2 + \left[z - R_1\sin\left(\frac{i-1}{4}\right)2\pi\right]^2} \tag{5.62}$$

Since the four lamps are symmetrical about the origin, the lowest point of the UV intensity in the area is at the origin.The minimum light intensity is:

$$I_{1\min} = 4 \times \frac{P\eta\eta_q}{2\pi R_1 L} e^{-\alpha_w (R_i - R_0)} \tag{5.63}$$

where:

α_w heree forabsorption coefficient of UVC, take 0.2326 cm^{-1};
R_0 rabsorption coefficient of U1.65 cm;
L UVC lamp effective length, according to the selected lamp parameters, 30 cm;
P UVC lamp input power, according to the selected lamp parameters, 3000 W;
η UV lamp conversion efficiency to UVC, according to the selected lamp parameters, 15.67%;
η_q conversion efficiency to UVC, take 0.9

Set $I_o = 400$ mW/cm^2, Plugging the above parameters into Eq. (5.63), the following is obtained:

$$R_1 = 7.13 \text{ cm}$$

Taking into account the assembly and optimization of light intensity, it is rounded to 7 cm.

② Calculation of the distribution radius of the outer ring

According to the second rule, points A and B are selected as the reference points of minimum light intensity. It is also assumed that the light intensity at points A and B is 400mW/cm^2, which is obtained according to Eq. (5.56):

$$I_A = \frac{p\eta\eta_q}{2\pi r_{AO3} L} e^{-\alpha_w (r_{AO3} - R_0)} + \frac{p\eta\eta_q}{2\pi r_{AO5} L} e^{-\alpha_w (r_{AO5} - R_0)} + \frac{p\eta\eta_q}{2\pi r_{AO6} L} e^{-\alpha_w (r_{AO6} - R_0)} + \frac{p\eta\eta_q}{2\pi r_{AO7} L} e^{-\alpha_w (r_{AO7} - R_0)} \tag{5.64}$$

$$I_B = \frac{p\eta\eta_q}{2\pi r_{BO3} L} e^{-\alpha_w (r_{BO3} - R_0)} + \frac{p\eta\eta_q}{2\pi r_{BO4} L} e^{-\alpha_w (r_{BO4} - R_0)} + \frac{p\eta\eta_q}{2\pi r_{BO1} L} e^{-\alpha_w (r_{BO1} - R_0)} + \frac{p\eta\eta_q}{2\pi r_{BO2} L} e^{-\alpha_w (r_{BO2} - R_0)} \tag{5.65}$$

Since the light intensity of two points A and B is the cumulative value of the four lamps in the vicinity, the four lamps are from the inner and outer rings. Therefore, assuming that the two lamps of the inner ring and the two lamps of the outer ring respectively provide an irradiation intensity of 200 mW/cm^2 to this point, the following can be obtained:

$$I_A = \frac{p\eta\eta_q}{2\pi r_{AO5} L} e^{-\alpha_w (r_{AO5} - R_0)} + \frac{p\eta\eta_q}{2\pi r_{AO6} L} e^{-\alpha_w (r_{AO6} - R_0)} = 200 \text{ mW/cm}^2 \tag{5.66}$$

$$I_B = \frac{p\eta\eta_q}{2\pi r_{BO3} L} e^{-\alpha_w (r_{BO3} - R_0)} + \frac{p\eta\eta_q}{2\pi r_{BO4} L} e^{-\alpha_w (r_{BO4} - R_0)} = 200 \text{ mW/cm}^2 \tag{5.67}$$

It can be seen from Fig. (5.17) that, $r_{AO5} = r_{AO6}$, $r_{BO3} = r_{BO4}$, Since the lamp distribution in the inner ring has been determined,plug $r_{BO3} = r_{BO4}$ into Eq. (5.66) to obtain the following:

$$r_{BO3} = r_{BO4} = 6.1847 \text{ cm}$$

Then according to Fig. 5.17, the length of the line segment BE can be calculated via trigonometric functions to be 5.35 cm. Its size is about 12.35 cm. Then take the 12.5 cm as the radius, draw a circle around the center, and optionally select two outer tube lamps to make the vertical bisector of the axis line of the two lamp tubes, which intersects the circle with radius R_2 at point A.Similarly, plug $r_{AO5} = r_{AO6}$ into Eq. (5.66), one can obtain $r_{AO5} = r_{AO6} = 6.1847$ cm. However, due to the relatively large distance between the outer ring lamps, the radial distance AD is less than BE, and the length of the AD calculated according to trigonometric functions is 3.87 cm. Therefore, the radius of R_3 is the sum of R_2 and AD, which is 16.22 cm, and is rounded to 16 cm.

(4) **Calculation of reactor inner wall radius**

According to the third regulation, the light intensity is the minimum point C. Since the sterilizer material is made of stainless steel with a high degree of finish, reflections should be taken into consideration when calculating the light intensity. According to general experience, the stainless steel reflection intensity is about 20%. Therefore, the total light intensity at point C is the incident intensity at point C plus the reflection intensity. which is:

$$I_C = 1.2 \times \left[\frac{p\eta\eta_q}{2\pi r_{CO1} L} e^{-\alpha_w (r_{CO1} - R_0)} + \frac{p\eta\eta_q}{2\pi r_{CO2} L} e^{-\alpha_w (r_{AO2} - R_0)} \right] \tag{5.68}$$

Since $r_{CO1} = r_{CO2}$, substituting into Eq. (5.68) to obtain the following:

$$r_{CO1} = r_{CO2} = 4.654 \text{ cm}$$

Similarly, the radial distance from point C to point R_3 is calculated to be 2.43 cm. In order to facilitate the sealing at both ends of the quartz tube, this dimension should be enlarged by 3 cm. That is, the inner radius R of the reactor is 19 cm.

In summary, the lamp distribution inside the reactor and the inner wall dimensions have been determined. By this size, the average irradiation intensity and the average dose of the reactor can be determined.

According to the special form of the target reactor lamp arrangement, the following method can be used to calculate its average light intensity.

The specific implementation method is to separate the two inner and outer lamp tubes, but their positions in the tube are not changed. The average optical intensities of the two rings are respectively calculated, and then the calculation results are added.

For the calculation of the average light intensity of the inner ring, since the average light intensity of the cylindrical sterilizer is equal to the average light intensity at any section along the lamp tube axis, taking x = C cross section, according to formula (5.61), the following can be derived:

$$\overline{I}_1(x) = \frac{1}{A}\iint\limits_{D}\sum_{i=1}^{4}\frac{P\eta\eta_q}{2\pi r_i L}e^{-\alpha_w(r_i - R_0)} \tag{5.69}$$

where:

A the area of the hollow part of the sterilizer, $A = \pi(R^2 - 14R_0^2)$;
r_i the distance from any point to the axis of the UV lamp,

$$r_i = \sqrt{\left[y - R_1\cos\left(\frac{i-1}{4}\right)2\pi\right]^2 + \left[z - R_1\sin\left(\frac{i-1}{4}\right)2\pi\right]^2};$$

R Sterilizer inner wall radius, 19 cm;
R_0 the radius of the quartz sleeve outer wall, 16.5 cm. Plug into formual (5.69), the following is obtained:

$$\overline{I}_1(x) = \frac{1}{\pi(R^2 - 14R_0^2)} \times \frac{P\eta\eta_q}{2\pi L}\iint\limits_{D}\sum_{i=1}^{4}\frac{e^{-\alpha_W\left(\sqrt{\left[y-R_1\cos\left(\frac{i-1}{4}\right)2\pi\right]^2+\left[z-R_1\sin\left(\frac{i-1}{4}\right)2\pi\right]^2}-R_0\right)}}{\sqrt{\left[y - R_1\cos\left(\frac{i-1}{4}\right)2\pi\right]^2 + \left[z - R_1\sin\left(\frac{i-1}{4}\right)2\pi\right]^2}}$$

The integration area is $y^2 + z^2 = 19^2$. Solving the above formula,the following is obtained:

$$\overline{I}_1(x) = 230.08\ \text{mW/cm}^2$$

In the same way, the average light intensity on the outer circumference can be calculated:

$$\overline{I}_2(x) = \frac{1}{\pi(R^2 - 14R_0^2)} \times \frac{P\eta\eta_q}{2\pi L}\iint\limits_{D}\sum_{i=1}^{10}\frac{e^{-\alpha_W\left(\sqrt{\left[y-R_3\cos\left(\frac{i-1}{4}\right)2\pi\right]^2+\left[z-R_3\sin\left(\frac{i-1}{4}\right)2\pi\right]^2}-R_0\right)}}{\sqrt{\left[y - R_3\cos\left(\frac{i-1}{4}\right)2\pi\right]^2 + \left[z - R_3\sin\left(\frac{i-1}{4}\right)2\pi\right]^2}}$$

The integration area is also the same, and the following can be obtained:

$$\overline{I}_2(x) = 575.2\ \text{mW/cm}^2$$

The average light intensity is:

$$\overline{I}(x) = \overline{I}_1(x) + \overline{I}_2(x) = 805.28\,\mathrm{mW/cm^2}$$

The calculation of the dose is inseparable from the average overcurrent time. The average overcurrent time is calculated as follows.

$$t = \frac{V}{Q} = \frac{\pi(R_2^2 - 14R_0^2)L}{Q} \tag{5.70}$$

where: $Q = 300\ \mathrm{m^3/h}$, $R_2 = 19$ cm, $R_0 = 1.65$ cm, $L = 30$ cm.

The above parameters can be plugged into Eq. (5.70)to get $t = 0.365$ s.

Therefore, the average dose is:

$$\overline{D} = \overline{I}(x) \times t = 805.28 \times 0.365 = 293.9\,\mathrm{mJ/cm^2}$$

Since the initial average dose requirement for the sterilizer is 250,000, and the target sterilizer is designed to arrange lamp tubes with an average dose of 293,900, the arrangement method meets the sterilization design requirements.

4. **Sterilizer structure design**

(1) **Sterilizer cylinder design**

The target of the sterilizer treatment is seawater. Therefore, the cylinder is made of stainless steel 316L with high chemical stability. Since the pressure of the ballast water in the pipeline will not exceed 0.6 MPa, 0.6 MPa may be used as the design pressure. The cylinder is a pressure vessel, and its wall thickness, flanges and heads are designed according to GB150 standard. The length of the sterilizer must be determined according to the length of the UV lamp, which is generally greater than the arc length of the UV lamp. The inner diameter is determined based on the design of the lamp distribution. The UVC illuminometer, inlet and outlet sampling ports, and exhaust pipes are designed and calculated according to standards. The structure is shown in Fig. 5.18.

(2) **Quartz sleeve end seal design**

Use the gland to press the "O" ring onto the cone. The end of the quartz sleeve is sealed and it has good centering. Its structure is shown in Fig. 5.19.

(3) **Casing cleaning device design**

When the UV lamp is in operation, the surface temperature of the lamp is high, and the quartz sleeve surface is easily fouled. Therefore, the cleaning device needs to be cleaned regularly. The structure of the cleaning device in this example is shown in Fig. 5.20. The cleaning rubber ring is matched with the sleeve under the action of a certain tension, and the wiper sleeve is realized by driving the cleaning plate through the screw. When the cleaning tray is in operation, a series of overflow holes are

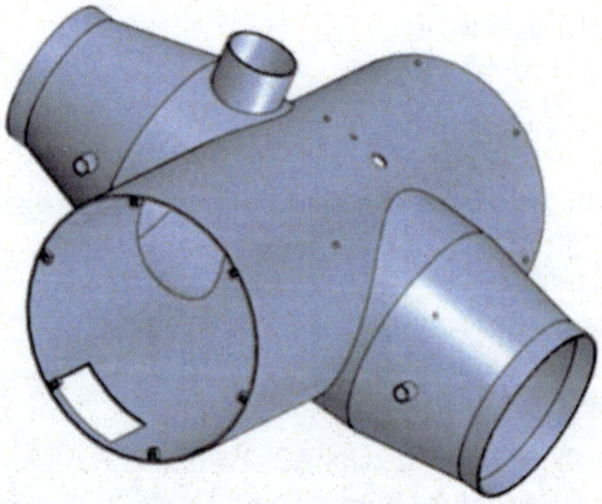

Fig. 5.18 Shell structure

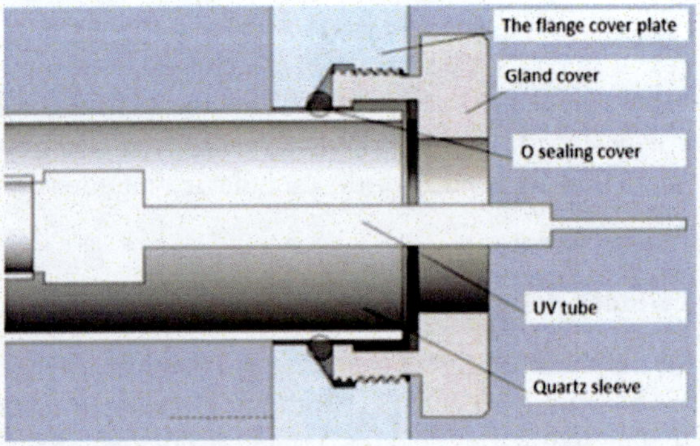

Fig. 5.19 Quartz sleeve end seal

Fig. 5.20 Cleaning device structure

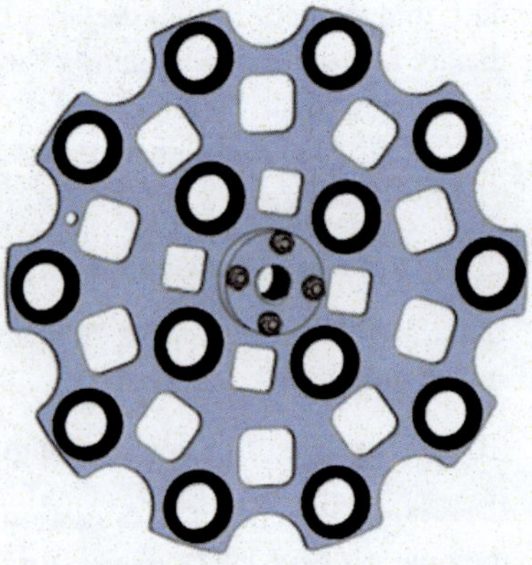

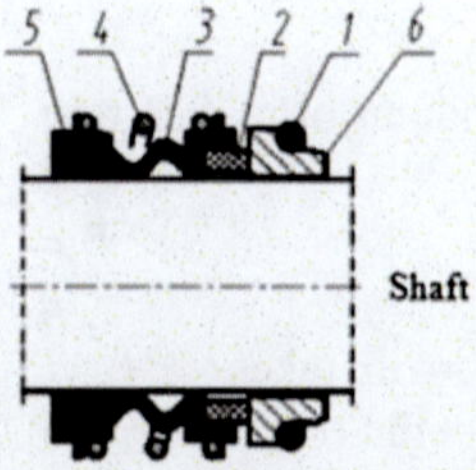

Fig. 5.21 Mechanical seal

1、Stationary ring，2、Swivel，3、Elastic element，
4、Spring，5、Spring seat，6、Auxiliary sealing ring

opened in the tray to reduce the resistance of the water flow. The limit switch is used to control the movement of the liquid crystals. The cleaning disk indirectly contacts the limit switch by means of a reciprocating touch guide rod, thereby realizing the expected movement mode.

(4) **Screw shaft seal design**

Since the number of revolutions of the screw mechanism in this device is very low, the form of the seal can be selected as a packing seal or a mechanical seal [15]. The resistance of the packing seal is larger than the mechanical seal, so the mechanical seal is chosen to achieve the seal effect. The specific structure is shown in Fig. 5.21.

(5) **Screw and gear motor design**

① Axial load calculation of screw

According to the working conditions, the following is obtained:

$$F = F_1 + F_2 + F_3 \tag{5.71}$$

where:

- F total axial load, N;
- F_1 The friction between the rubber ring and the quartz tube caused by the initial compression, N;
- F_2 The friction between the rubber ring and the quartz tube caused by water pressure, N;
- F_3 The resistance of the flow to the wash plate, N.

I. The friction caused by initial compression on each quartz sleeve

According to the following formula:

$$F_1' = \frac{0.2\pi^2 feEDW}{1 - \mu^2} \tag{5.72}$$

where:

E Elastic modulus of rubber, take

$$E = 0.0078 \times 10^9 GPa$$

μ Poisson's ratio, take $\mu = 0.47$;
f friction coefficient, take $f = 0.75$;
W The thickness of the contact portion of the rubber ring, which is available according to the structural dimensions of the rubber ring: $W = 2$ mm;
D The outer diameter of the rubber ring contact part, which is available according to the structural dimensions of the rubber ring: $D = 38$ mm;
e Compression ratio, available according to the structural dimensions of the rubber ring: $e = 0.1$.

Substituting the data above, the following is obtained:

$$F_1' = \frac{0.2 \times 3.14^2 \times 0.75 \times 0.1 \times 0.0078 \times 10^9 \times 38 \times 10^{-3} \times 2 \times 10^{-3}}{1 - (0.47)^2} \approx 113N$$

Since the device has 14 tubes with the same parameters, the following can be derived: F1 $= 14 \times F_1' = 14 \times 113 = 1576$ N.

II. Friction between rubber ring and quartz sleeve caused by water pressure

First, the pressure difference between the two ends of the cleaning plate should be calculated. According to the size of the cleaning plate, the total area A_0 is 109804 mm^2, the physical area is $A_1 = 63132$ mm^2, and the area of the lamp hole that leaks out is $A_3 = 12704.4$ mm^2. Therefore, the area is $A_2 = A_0 - A_1 - A_3 = 33967.6$ mm^2.

Due to traffic:

$$Q = 300\ \mathrm{m^3/h} = \frac{1}{12}\ \mathrm{m^3}/s$$

Then for the total area of the flow rate:

$$V_1 = \frac{Q}{A_0} = \frac{1/12}{109804 \times 10^{-6}} = 0.758928\ \mathrm{m/s}$$

Similarly, for the flow rate of the empty area:

$$V_2 = \frac{Q}{A_2} = 2.45332\ \mathrm{m/s}$$

Regardless of the friction loss, taking the horizontaldirection to the right as the positive direction, the following can be obtained according tothe momentum theorem:

$$F_y = \rho Q(V_2 - V_1) = 10^3 \times \frac{1}{12}(2.45332 - 0.5242326) \approx 161 \text{ N}$$

Therefore, the pressure difference is:

$$P = \frac{F_y}{A_1} = \frac{161}{63132 \times 10^{-6}} \approx 2250 \text{ Pa}$$

Based on what we obtained above, the friction between a single rubber ring and a single quartz tube caused by water pressure is calculated as follows.

$$F_2' = \frac{f\pi(1+\mu)PDW}{1-\mu^2} = \frac{0.75 \times 3.14 \times (1+0.47) \times 2250 \times 38 \times 10^{-3} \times 2 \times 10^{-3}}{1-(0.47)^2} \approx 0.8 \text{ N}$$

Since the device has the 14 lamps with same parameters, the following can be derived: F2 $= 14 \times F_2' = 14 \times 0.8 = 11.2$ N.

III. The calculation of the resistance of the water flow to the cleaning disk.

From the calculation in the second part, it is precisely because the resistance of the water flow causes a pressure difference across the cleaning plate, so the flow resistance is:

$$F_3 = F_y = 161 \text{ N}$$

In summary:

$$F = F_1 + F_2 + F_3 = 1748.2 \text{ N}$$

which is rounded to 1750 N.

② Screw design and calculation:

According to

$$d_2 = \zeta\sqrt{\frac{F}{\phi[P]}} \tag{5.73}$$

where:

F the axial load of the screw (N);
$[P]$ permissible pressure (MPa);

The value of φ can be selected according to the form of the nut, the integral nut = 1.2 −2.5 andthe split nut = 2.5−3.5.

From the tablebased on the material, take $[P]$= 10, φ= 2. The lead screw uses trapezoidal thread. So, ζ = 0.8. Plugging into Eq. (5.73), the following is obtained:

$$d_2 = 0.8 \times \sqrt{\frac{1750}{2 \times 10}} = 7.5 \text{ mm}$$

Taking into account the length of the screw, stiffness, deflection and the size of the screw shaft at both ends and after looking up the reference GB/T 5796.2-2005, we take the screw with P = 5 mm, d = 28 mm.

Calculation of equivalent friction angle ρ':

$$\rho' = \arctan\frac{\mu_s}{\cos\frac{\alpha}{2}} \tag{5.74}$$

As the screw is selected to be trapezoidal thread, the angle of the tooth profile $\alpha = 30°$, the friction coefficient of the spiral pair is looked up to be $\mu_s = 0.15$. Substituting the data into Eq. (5.74), we have:

$$\rho' = \arctan\frac{\mu_s}{\cos\frac{\alpha}{2}} = \arctan\frac{0.15}{\cos 15^\circ} = 8.83^\circ$$

Calculation of thread angle ψ:

$$\psi = \arctan\frac{L}{\pi d_2} \tag{5.75}$$

where:

L Lead, the lead screw is a single lead screw, the lead is the pitch, that is L = 5 mm.

d_2 The center of the screw, $d_2 = 25.5$ mm.

Substituting data above into formula (5.75) yields:

$$\psi = \arctan\frac{L}{\pi d_2} = \arctan\frac{5}{3.14 \times 25.5} = 3.57^\circ$$

The calculation shows that the screw is self-locking.

③ Total drive torque calculation:

For thread torque calculations:

$$T_1 = F\frac{d_2}{2}\tan(\psi + \rho') \tag{5.76}$$

where:

F axial load, $F = 1750$ N;

d_2 The diameter of the screw, $d_2 = 25.5$ mm;

ψ Thread angle, $\psi = 3.57°$;

ρ' friction angle, $\rho' = 8.83°$.

Substituting the data above into formula (5.76) yields:

$$T_1 = F\frac{d_2}{2}\tan(\psi + \rho') = 1750 \times \frac{25.5 \times 10^{-3}}{2}\tan(8.83^{\circ} + 3.57^{\circ}) = 4.91N \cdot m$$

Screw efficiency η_1 is,

$$\eta_1 = \frac{\tan\psi}{\tan(\psi \pm \rho')} \tag{5.77}$$

The number is taken when the axial load is opposite to the direction of motion. Substitution the data above into formual (5.77) yields:

$$\eta_1 = \frac{\tan\psi}{\tan(\psi \pm \rho')} = \frac{\tan 3.57^{\circ}}{\tan(3.57^{\circ} + 8.83^{\circ})} = 0.284$$

Angular Contact Bearing Efficiency $\eta_2 = 0.95$, Mechanical Efficiency of Elastic Couplings $\eta_3 = 0.99$.

For the calculation of mechanical seal face friction torque T_f:

$$T_f = \frac{P_f}{\omega} = f\pi d_m b p_c \frac{d_1 + d_2}{4} \tag{5.78}$$

where:

d_m The average diameter (m) of the sealing face, $d_m = \frac{d_1+d_2}{2}$;
b Seal end face width (m), $b = \frac{d_2-d_1}{2}$
f Face friction coefficient, here $f = 0.1$
p_c Sealed end face pressure (Pa), $p_c = 0.4 \times 10^6$ Pa
d_1 Inner diameter (m) of the seal face, $d_1 = 0.028$ m
d_2 Sealed outside diameter (m), $d_2 = 0.034$ m

Substituting the data above into formula (5.78) yields:

$$T_f = 0.1 \times 3.14 \times \frac{0.034 + 0.028}{2} \times \frac{0.034 - 0.028}{2} \times 0.4 \times 10^6 \times \frac{0.034 + 0.028}{4}$$
$$= 0.181N \cdot m$$

Total drive torque T calculation is,

$$T = \frac{T_1}{\eta_1 \cdot \eta_2 \cdot \eta_3} + 2T_f = \frac{4.91}{0.284 \times 0.95 \times 0.99} + 2 \times 0.181 = 18.8N \cdot m$$

In summary, the screw profile of the designed screw is a trapezoidal thread with a nominal diameter of 28 mm.The selected geared motor model is NMRV40, the motor power is 250 W, and the output torque and the number of revolutions are 36 N·m and 47r/min, respectively. The models are shown in Figs. 5.22 and 5.23, respectively.

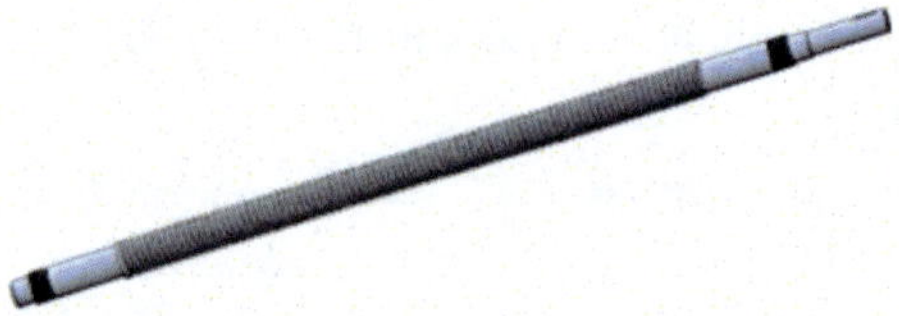

Fig. 5.22 Screw

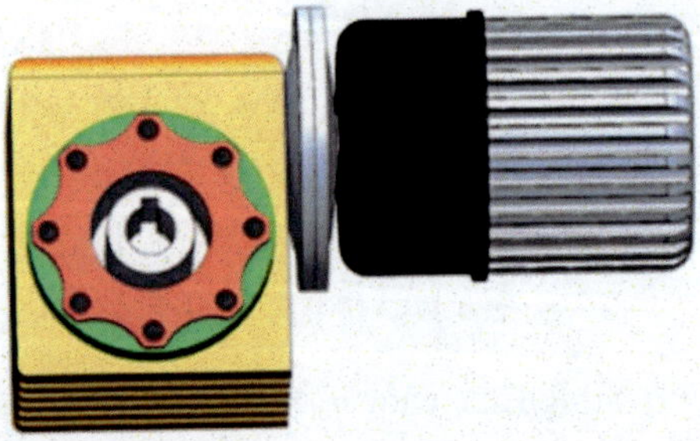

Fig. 5.23 Deceleration motor model

After the above design calculations, the UVC sterilizer design was finalized. The specific parameters are as follows.

Processing flow: Q = 300 m^3;
Use of materials: 316L stainless steel;
Sterilizer inner diameter: d = 380 mm;
The total length of sterilizer: L = 796 mm;
Internal barrel speed: v = 0.735 m/s;
Lamp number: 14;
Import and export diameter: d2 = 211 mm,
Import and export flow rate: v = 2.38 m/s.
Maximum working temperature: 80 °C

The specific structure is shown in Fig. 5.24.

Fig. 5.24 Drawing of Uvc general assembly and drawing of exploded

5.3.2 *Design of Inert Gas Sterilization Unit*

Inert gas sterilisation, also known as scavenging technology, is based on the rule that any microorganism is difficult to survive when leaving oxygen, using pure nitrogen or an inert gas to drive off the oxygen in the water in order to inhibit bacterial growth. The long-range ship can prevent the re-growth of microorganisms. At the same time, the inert gas can also prevent the corrosion of the bulkhead and reduce the power consumption of cleaning.

Because the inert gas sterilization technology injects a large amount of N_2 gas into the ballast water according to the principles of the partial pressure law and the total pressure law, the O_2 dissolved in the ballast water is driven away from the seawater to achieve the purpose of inhibiting the growth of bacteria. The inert gas sterilization unit becomes the entire ballast tank, so long as a nitrogen generator or an inert gas generator is purchased and the generated N_2 or combustion-cleaned inert gas is injected into the ballast tank, killing and inhibiting microorganisms can be achieved. For the purpose of propagation in ballast tanks, the injection and diffusion processes of inert gas into the tank are shown in Figs. 5.25 and 5.26. Therefore, to maintain the pressure in the cabin, to achieve sterilization of the entire ballast tank, and to inhibit the growth of bacteria, a breather valve is required at the top of the ballast tank to achieve the function of sterilization of the ballast tank.

In order to maintain the pressure in the tank, the breathing valve installed on the air tube head at the top of the ballast tank is also called P/V valve, as shown in Fig. 5.27.

The role of the P/V valve is: (1) to keep the ballast compartment closed and slightly positive pressure to ensure that under this pressure, the partial pressure of the inert gas is greater than the dissolved O2 in the ballast water, allowing the dissolved oxygen in the ballast water to overflow; (2) to prevent seawater fromentering the ballast tank from the outside; (3) When the air or seawater enters the ballast tank, it prevents excessive pressure or vacuum in the ballast tank; (4) Automatic drainage; (5) It can work normally when the ship is tilted ± 40°. As shown in Fig. 4.23, the

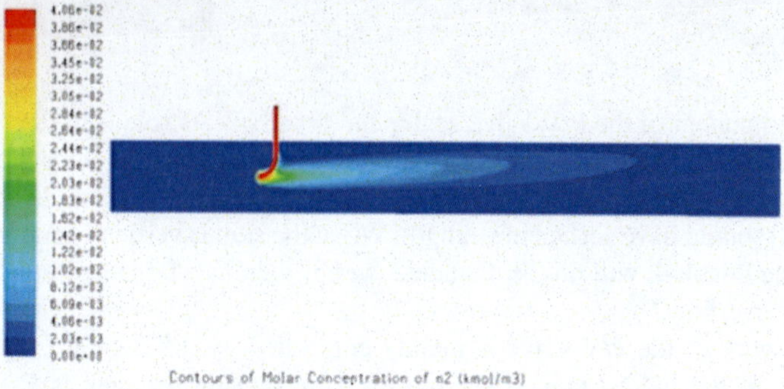

Fig. 5.25 Inert gas injection simulation

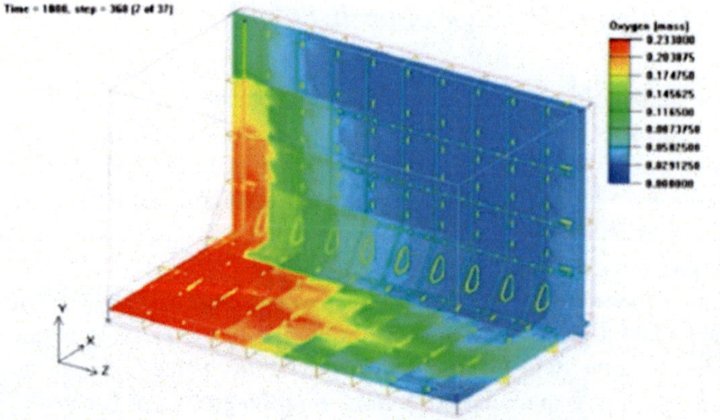

Fig. 5.26 Inert gas in ballast cabin diffusion

Fig. 5.27 P/V valve

main component of the P/V valve seal is the float ball. Its role is to float the floated ball of the lower valve disc when the seawater is poured back, blocking the upper port and preventing the seawater from pouring back into the ballast tank. In addition, the float ball should have sufficient strength. When the ship is vertically and horizontally tilted, the float ball will not be damaged even if there is a large hydraulic impact on the tank overflow.

The work of the P/V valve is mainly controlled by differential pressure. If the pressure in the ballast tank is P_1, the pressure outside the tank is P_2. When the internal and external pressure is balanced, that is, $P_1 = P_2$, the P/V valve upper and lower valve plates are closed; When loading, the pressure in the tank is greater than the pressure outside the tank, i.e., when $P_1 > P_2$, the lower valve disk is lifted up. The O_2 and N_2 discharged under the partial pressure are discharged from the tank and release the pressure so that the ballast can be performed normally. Generally, the pressure difference is between 0.002 and 0.003 MPa; When the vessel is unloaded, the pressure in the tank is lower than the pressure outside the tank, i.e., when $P_1 < P_2$, the upper valve plate is lifted up, so that the ballast tank is open to the external atmosphere, and the outside atmosphere enters the ballast tank, so that the unloading can be performed normally. In general, the pressure difference is between −0.002 and −0.003 MPa.

Compared with air fittings, the flow rate calculation is required when designing and selecting a ballast air venting system. According to the revised version of the IACS standard 1991 and 2012, the wall thickness of the inner and outer chambers of the automatic closing air tube head shall not be less than 6 mm in accordance with the requirement of the air tube head closing device P_3 (P_3 Air Pipe Clossing Devices). For the reliability of the designed P/V valve air pipe joints, a flow limit test is to be carried out to prevent the valve head structure from being implicated in the displacement of the upper and lower valve discs and the float ball, which limits the free movement of the relevant moving parts under different working conditions.

Taking the P/V valve design of the DN200 ballast tank vent pipe as an example, the maximum flow rate of the P/V valve was measured by experiments as $Q = 380\ m^3/h$. The limiting flow rate was calculated to be 3.36 m/s viaequation $V_R = 4Q \times 10^6/3600 \times DN2$. Considering the air diffusion after flowing into the atmosphere, the maximum air velocity $V_{max} = 80\% V_R = 2.69$ m/s. For DN250 breathers, the P/V valve has a maximum air flow VMax = 3.4 m/s. Other permeable tubes can be found in the relevant classification societies test report. Generally, in the flow state, the additional fluid resistance of the P/V valve, either in or out, is approximately 300 mm–H_2O high, which satisfies the pressure range that the ballast tank can withstand.

5.3.3 Photocatalytic Sterilization Unit Design

The design of the photoelectrocatalytic sterilization unit is basically the same as the design method of the UVC sterilization unit. The difference between the two is the elimination of the mechanical cleaning mechanism of the UVC sterilization unit from the photocatalytic sterilization unit. In addition, no US cleaning device is added on both sides of the UVC lamp. Instead, around the quartz sheath of the UVC lamp, a layer of transparent nano-TiO_2 coating is sintered on the surface of the coating. Under UVC irradiation, the coating forms OH radicals with water. Itssystem working process is as shown in the Fig. 5.28.

When nano-TiO_2 receives photon radiation ($\leq$ 388 nm) with energy greater than or equal to its own forbidden band width, the valence band electrons of nano-TiO_2 are excited to the conduction band, resulting in photo-generated electron-hole pairs. Under the action of an electric field, electron-hole pairs separate and migrate to the particle surface. Photovoltaic holes carry the main part of the photon energy and

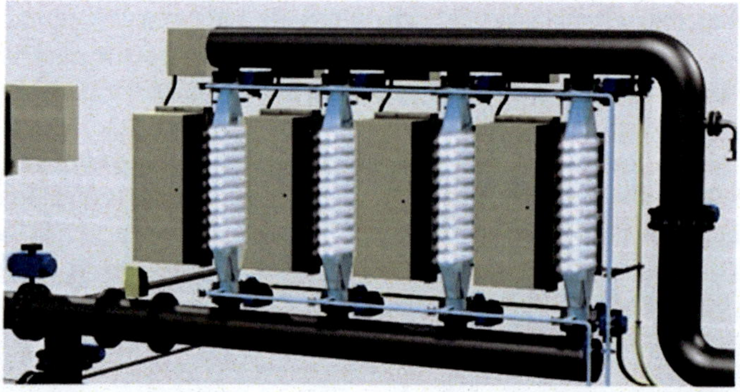

Fig. 5.28 Photoelectric catalysis of the unit

have extremely strong electron-harvesting ability. They react with OH− and H_2O molecules adsorbed on the surface to generate hydroxyl radicals and OH with strong oxidative properties. In the process of the dissolution of hydroxyl radicals, a series of free radical chain reactions will be induced to oxidize and decompose organic substances and microorganisms, and finally they will be degraded into CO_2, H_2O and trace inorganic salts, and there will be no toxic and toxic residues. The basic reaction formula can be expressed as follows [16, 17].

$$TiO_2 + hv \rightarrow TiO_2 + h^+ + e^-$$

$$H_2O \rightarrow OH^- + H^+$$

$$h^+ + OH^- \rightarrow \bullet OH$$

$$h^+ + H_2O + O_2^- \rightarrow \bullet OH + H^+ + \bullet O_2^-$$

$$h^+ + H_2O \rightarrow \bullet OH + H^+$$

$$e^- + O_2 \rightarrow O_2^-$$

$$\bullet O_2^- + H^+ \rightarrow \bullet HO_2$$

$$2 \bullet HO_2 \rightarrow H_2O_2 + O_2$$

$$H_2O_2 + \bullet O_2^- \rightarrow \bullet OH^+ OH^- + O_2$$

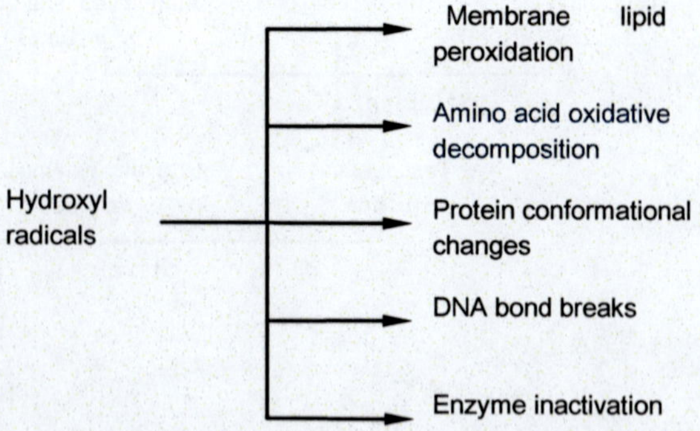

Fig. 5.29 The hydroxyl free radical mechanism to kill microorganisms

$$H_2O_2 + hv \rightarrow 2 \bullet OH$$

$$\text{Organ} + \bullet OH + O_2 \rightarrow CO_2 + H_2O$$

The mechanism of microbial killing by hydroxyl radicals is shown in Fig. 5.29.

Therefore, under the combined action of UVC and its photocatalysis, a good bactericidal effect can be achieved, and a pure quartz tube nano-TiO_2 photocatalytic film is prepared.

Preparation of nano-coating methods include chemical vapor deposition, sputtering, sol-gel method, hydrothermal method, adhesive method, electrical deposition method, pulsed laser deposition method and liquid deposition method [18]. The sol-gel method has the advantages of low cost, easy control, low synthesis temperature, low requirement on the reaction system process conditions, high product purity, small particle size, and uniform composition. Based on the above considerations, the sol-gel method was used to prepare nano-TiO_2 sols. The nano-TiO_2 coating was deposited on a pure quartz tube by spray drying method, and then the sintering method was used to make the nano-TiO_2 coating sintered and solidified on the tube of pure quartz.

The sol-gel process can be divided into two methods based on raw materials and types: organic metal alkoxide method and inorganic salt method. The hydrolysis and polycondensation reaction of metal organic alkoxides can be expressed as:

Hydrolysis reaction: $M(OR)_4 + nH_2O \rightarrow M(OR)_{4\text{-}n}(OH)_n + nROH$

Polycondensation reaction: $2\ M(OR)_{4-n}\ (OH)n \rightarrow [M(OR)_{4-n}\ (OH)_{n-1}]_2O + H_2O$ where: M representing metal,R represents alkyl.

The total response can be expressed as: $M(OR)_4 + 2H_2O \rightarrow MO_2 + 4ROH$

The preparation process of TiO_2 photocatalytic film [19, 20] is as follows. Preparation of 0.5 mol/L nano-TiO_2 sol: The precursor was prepared using tetrabutyl titanate as nano-TiO_2, ethanol and distilled water as diluents, acetylacetone as complexing agent, and nitric acid as hydrolysis catalyst. The photocatalytic nano-film preparation process is shown in Fig. 5.30. First, a certain volume of tetrabutyl titanate and diethyl acetonate are precisely measured by a pipette and added to one-half of the prepared ethanol. Stir the stirrer for 5 min and mix well. The remaining one-half of

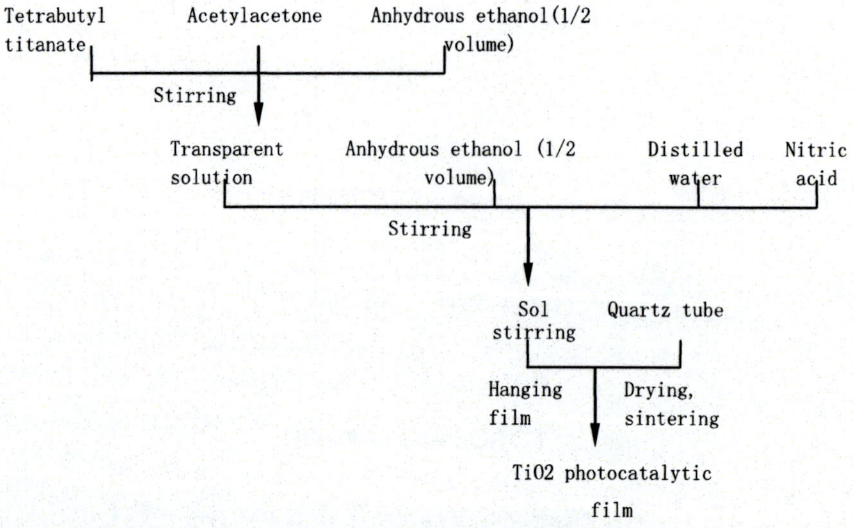

Fig. 5.30 Nanometer TiO_2 photocatalytic membrane preparation process

absolute ethanol is then added with stirring and the appropriate amount of distilled water and nitric acid are added in proportion. The stirring is continued until the sol is clear and transparent, and the sol is then closed for use. A nanometer TiO_2 sol was prepared at a concentration of 0.5 mol/L. The amount of each component in the sol was as follows.

tetrabutyl titanate: absolute ethanol: acetylacetone: distilled water: nitric acid = 1.0:26.5:1.0:1.0:0.2.

Preparation of Nano-TiO_2 Pure Quartz Tubes: First, a pure quartz tube is used as a carrier for the film, and the tube is processed before the film is formed. First wash the surface dust with water, and then use the detergent to wash away the organic substances that remain on the surface, after thatsoak it in diluted hydrochloric acid for 10 min, rinse it with distilled water, and dry it for later use. Secondly, the cleaned and dried pure quartz tube is erected and sealed. The prepared sol is sprayed onto a pure quartz tube uniformly with a watering can, dried at room temperature, and placed in a muffle furnace for roasting. The temperature was raised to 500 °C at a rate of 2 °C/min, and the temperature was maintained for 1 h. The furnace was cooled. A transparent TiO_2 photocatalytic film can be obtained on the outer surface of the quartz tube.

In order to ensure the ultraviolet transmittance of pure silica glass tube, only one layer of nano-TiO_2 film should be deposited on the pure silica glass tube carrier. This layer of film has little effect on the penetration of ultraviolet light on pure quartz glass tubes, and only colored refracted light is looming in some places. According to the literature [21], scanning the outer surface of the quartz tube by atomic force microscope can obtain the thin-film electron micrograph (AFM diagram) as shown in Fig. 5.31. The resulting nano-TiO_2 film is tightly bound to the quartz glass carrier

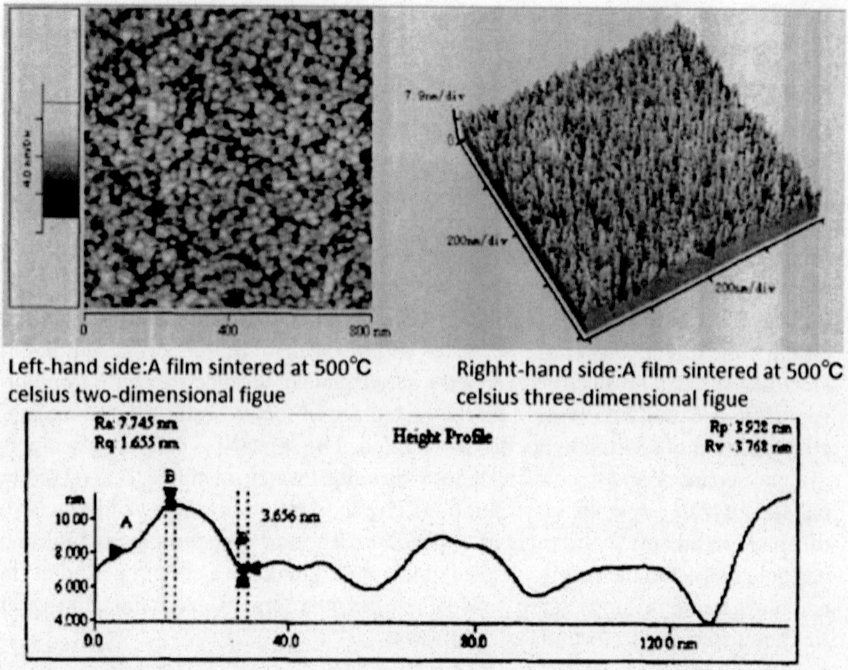

Fig. 5.31 A layer of nanometer TiO_2 thin film electron microscope scanning figure

and is an undulating granular film. The particles are uniformly dense, whicharesmall, about 20–30 nm, and the grain size is in the nanometer range. With the decrease of particle size, nanometer TiO_2 rapidly increases the number of surface atoms, improves the light absorption efficiency, increases the concentration of photogenerated carriers on the surface, increases the number of redox reactions, and accelerates the photocatalytic reaction rate, with higher photocatalytic activity.

The hydroxylradical·OH generated by photoelectrocatalysis is only 10 ns in the natural state. Therefore, the oxidation performance of the hydroxyl radical· OH is not corrosive to the ballast piping of the vessel except that it is sterilized with UVC in the photocatalytic sterilization unit. At the same time, since the sterilizing unit is designed to eliminate the US cleaning device on both sides of the mechanical cleaning mechanism and the UVC lamp, after the system processes the ballast water, the formic acid must be used for cleaning to prevent the quartz sleeve surface from being damaged. The fouling caused by the high temperature of the UVC lamp ensures the transmittance of the quartz tube.

5.3.4 *Design of Electrolysis or Potion Sterilization Units*

Both the electrolysis method and the sterilizing unit use the hypochlorous acid as an effective inactivator to kill the plankton, pathogens, larvae or spores that remain in the ballasted seawater after filtration.

1. **Electrolytic method**

The core component of electrolysis is the tube plate type electrolyzer. The filtered seawater will be diverted into the tube plate type electrolyzer by about 1–2% all the way through the filtered seawater. When both ends of the electrolysis tank are on the same direct current, the seawater will be electrolyzed under the action of the electrode and will be electrolyzed on the anode plate of the electrolyzer. The chlorine gas is ionized and the chlorine dissolves rapidly in the seawater to produce a high concentration of sodium hypochlorite solution. The solution is dehydrogenated by a cyclone separator and injected back into the ballast water main line. It is mixed with ballast water to a specific concentration. The sodium hypochlorite solution can kill the microorganisms that remain after filtration. The active sodium hypochlorite that has not completed the reaction exists in the ballast tank for a certain period of time, and plays a role in suppressing residual marine microorganisms in the ballast water. Its chemical reaction formula is as follows.

$$\text{anode: } 2Cl^- \rightarrow Cl_2 + 2e$$

$$\text{cathode: } 2H_2O + 2e \rightarrow 2OH^- + H_2 \uparrow$$

Chlorine can dissolve in water to produce hypochlorous acid and hydrochloric acid : $Cl_2 + H_2O \rightarrow HOCl + Cl^- + H^-$

$$\text{Total response: } NaCl + H_2O \rightarrow NaOCl + H_2 \uparrow$$

The reaction process of the system is shown in Fig. 5.32.

The hydrogen separated in the cyclone separator is separated from the water/water separator at the top of the cyclone separator and then enters the reflux chamber. The gas is then discharged from the top diversion orifice into the hydrogen discharge line. The two high-power fans in the system inject air to dilute the hydrogen concentration to a safe range and discharge it to a safe area on the deck, as shown in Fig. 5.33.

When unloading, the ballast water does not need to go through the electrolysis unit and the filter, but directly outboard, and the residual oxidant concentrationTRO sensor installed at the outboard opening of the ballast water monitors the TRO concentration of the discharged ballast water. If the TRO concentration is higher than 0.1 ppm, the system will automatically start the neutralization unit and inject the neutralizing agent sodium thiosulfate ($Na_2S_2O_3$) into the row of outboard ballast water to neutralize the residual oxidant sodium hypochlorite. The principle of neutralization is:

Fig. 5.32 System reaction process

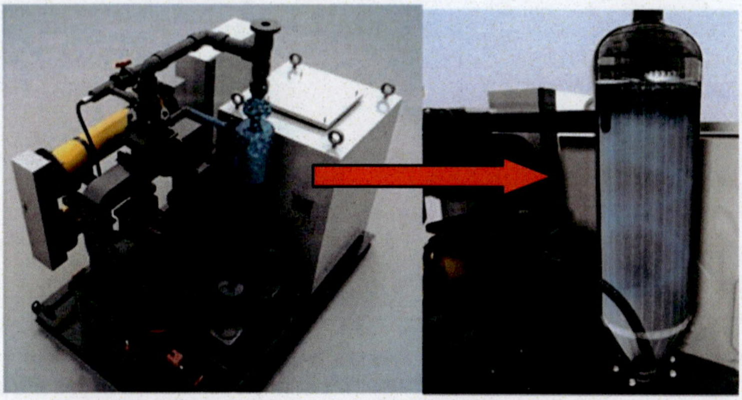

Fig. 5.33 H_2expulsion process

$$Na_2S_2O_3 + 4NaClO + H_2O \rightarrow 2NaHSO_4 + 4NaCl$$

The amount of neutralizing agent is: Neutralizer adjusted to 3−7 kg $Na_2S_2O_3$ dry powder per 1000 m^3 ballast water. The neutralizer is injected by a diaphragm metering pump.

If the TRO concentration is less than 0.1 ppm, the integrated unit will not start and the ballast water will be directly discharged to the destination sea area.

Plate-plate electrolyzers are designed according to the standard GB 12176-90 [22]. The ability of the sodium hypochlorite solution to oxidize is quantified by the available chlorine concentration (C) to indicate the oxidizing power per liter of solution, which corresponds to the oxidizing power of several grams of chlorine in water. The unit is g/L. The available chlorine concentration is equal to twice the concentration of chlorine in the positive state of the solution. For every 1g of sodium hypochlorite contained in the solution, 0.953g of available chlorine was contained.

The yield of oxidant sodium hypochlorite in the tube-plate electrolyzer is expressed by the available chlorine yield, which is equal to the mass (g) unit g/h of chlorine produced per hour when the equipment is working under rated conditions. The effective chlorine yield is calculated according to the following formula:

$$G = C \times Q \tag{5.79}$$

where:Q- the flow of sodium hypochlorite solution per hour, L/h.

The ratio of the actual production of available chlorine to the theoretical production after a certain amount of current flows through the tube-plate electrolytic cell is referred to as the current efficiency (η) of the cell. According to Faraday's law of electrolysis, the theoretical production of available chlorine is 1.323g per 1A·h of current flow in the electrolyzer. The current efficiency is calculated according to the following formula (5.80):

$$\eta = G/I \times n \times 1.323 \times 100\% \tag{5.80}$$

where:

I	Electrolytic current, A;
n	Electrode series;
1.323	Theoretical production of current effective chlorine per ampere-hour, g/A·h.

When a tube-plate type electrolytic cell is operated in a rated state, a direct voltage applied between the anode and the anode of the electrolytic cell is called an electrolytic voltage (V). When the electrolytic cell is operated by a plurality of pairs of cathode and anode series power supply, the electrolysis voltage is expressed by multiplying the electrolysis voltage between each pair of anode and cathode by the series stage. such as 4 V × 3.

When the tube plate type electrolytic cell maintains the rated yield, the value of the electrolytic current flowing in the electrolytic cell is referred to as the rated electrolytic current (I). When the electrolytic cell uses multiple pairs of cathodes and anodes to supply power in parallel, the rated electrolytic current can be represented by multiplying the current between each pair of electrodes and the number of parallel poles, such as 50 A × 2.

The tube plate type electrolyzer uses seawater as the electrolyte. The electrolyte concentration (S) is expressed in grams of NaCl per liter of solution (g/L).

When the tube-plate electrolyzer is working under rated conditions, the DC energy consumed for generating 1 kg of available chlorine in the electrolyzer is called the DC power consumption (PDC), and the unit is (kW·h)/kg. The formula for calculation is shown in (5.81) below:

$$PDC = U \times I/G = U \times I/Q \times C \tag{5.81}$$

where:

U Electrolysis voltage (VDC);
I Electrolysis current (ADC);
G Available chlorine yield (g/h);
Q Production of secondary chlorine solution (L/h);
C Available chlorine concentration of sodium hypochlorite (g/L)。

Generally, the effective chlorine yield of the tube plate type electrolysis tank for ship ballast water needs to be calculated by using 15−30 kg for the ballast water of 1000 m^3.

2. **Pharmacy method**

The principle of the pharmacy method of killing marine microorganisms is the same as that of the electrolysis method. The difference is that the pharmacy method changes sodium hypochlorite generated from electrolysis of seawater in a tube plate type electrolyzer into sodium hypochlorite aqueous solution prepared directly using sodium hypochlorite powder.

When the ship is ballasted, the active material of sodium hypochlorite aqueous solution with an effective chlorine yield of 15−30 kg is reversely injected by a diaphragm pump with 1000 m^3 of ballast water to kill the residual marine microorganisms after filtration. In this process, since there is no production of electrolytic H_2, the system is simplified and the safety of the system is also ensured.

When unloading, the working process is the same as that of the electrolysis method. It is also based on the active substance concentration detected by the TRO sensor. The neutralizing agent sodium thiosulfate ($Na_2S_2O_3$) is used to neutralize residual oxidant sodium hypochlorite.

References

1. Ai-wei, Wang. 2008. *Development and theoretical research of sucking backwash filter*. Beijing University of Chemical Technology, pp 6-7.
2. Ai-wei, Wang, Zhu Hai-jun, Qian Cai-fu et al. 2007. Back flush filter structure characteristics and application prospects. *Petrochemical Equipment* 68–69.
3. Li-fu, Tang, Wang Wei-yi, and Zhang Huai-qing. 1984. *Filter*, 10–12. Beijing: Machinery Industry Press.
4. Ol, Clyde and Shao Qiyao (Translation). 1982. *Filtering Theory and Practice*. Beijing, National Defense Industry Press, pp 17-18.
5. Huicheng, Zhou and Congsheng. 1997. Development and application of backwash filters at home and abroad. *Technology* (12):32-33.
6. Qi, Wen. 2004. Research on filter mechanism analysis and control system of automatic backwash filter.*Journal of Zhejiang University* 39-40.
7. Ruina, Chen. 2008. Design and mechanism of ultrasonic automatic backwash filter. Dalian Jiaotong University Master Thesis, pp. 41-44.
8. Xueshi, Pang. 2010. *Hydrocyclone Technology and Application*, 264–275. Beijing: Sinopec Publishing House.

9. Silva, D.O., L.G.M. Vieira, and M.A.S. Barrozo. 2015. Optimization of Design and Performance of Solid-Liquid Separators: A ThickenerHydrocyclone[J]. *Chemical Engineering and Technology* 38 (2): 319–326.
10. Zhang, J., X.Y. You, and Z.G. Niu. 2011. Numerical simulation of solid-liquid flow in hydrocyclone. *Chemical and Biochemical Engineering Quarterly* 25(1): 37-41.
11. Qi-nian, Yan. 2011. Research and development of broadband ultrasonic sewage treatment device. *Master's Degree Thesisof Hebei University* 5: 17–20.
12. Rui-na, Chen. 2008. Design and Mechanism of Ultrasonic Automatic Backwash Filter. *Dissertation of Dalian Jiaotong University* 12: 21–29.
13. Wang Hai-ou, Hu, Wang Jian-nan Zhi-chao, et al. 2011. Effects of connection mode between transducer and tank on ultrasonic field distribution. *Transactions of the Chinese Society of Agricultural Engineering* 27(2): 26–30.
14. Bogdan, N. 2009. Influence of concentration of substances used in ultrasonic cleaning in alkaline solutions on cavitation intensity. *Ultrasonics Sonochemistry* 16(3): 402–407.
15. Wang, B. and A.B. Yu. 2008. Numberical study of the gas-liquid-solid flow in hdrocyclone with different configuration of vortex finder. *Chemical Engineering Journal* 135: 33-42.
16. Chun-mei, Wang. 2005. The property of titanium dioxide photocatalyst and its applicationto the water treatment.*Journal of Nantong Institute of Technology(Natural Science)* 4(1): 28-29.
17. Xin, Xin and Li Zhao-bo. 2003. Photocatalytic performance of nanometer TiO_2 and its application in water treatment.*Anhui Chemical Industry* (6): 32-35.
18. Meng, Sun. 2005. Preparation and photocatalytic properties of nano-TiO2 films. Zhengzhou: Master's thesis, Zhengzhou University, China, pp. 8-12.
19. Kai-shu, Guan, Bai Hao-ran and YIN Yan-sheng. 2003. Preparation and photocatalytic behavior of tio2coatings on glass. *Mechanical Engineering Materials* 27(2): 34-35.
20. Cheng-yan, Wang, Liu Xi-min and Ran Xiang-cheng et al. 2009. Preparation of nanosized TiO_2 and its applied in COD detection. *China Environmental Testing* 25(3): 66-6.
21. Ding-jian, Li. 2004. Application of nano-TiO2 photocatalytic sterilization technology in water body and discussion on its bactericidal mechanism. Guangzhou: Jinan University, pp. 12-20.
22. National Standard of the People's Republic of China. 1990. Sodium Hypochlorite Generator GB12176–90. Beijing: State Bureau of Technical Supervision.

Chapter 6
Optimization of Key Components

6.1 Filtering Equipment Optimization

As a key component for ballast water pretreatment, whether it is a backwash filter or a swirl filter, it plays an indispensable role in the ballast water biological control system. The backwash filter has high filtration accuracy, but it must be continuously cleaned in order to reduce the pressure loss and flow rate reduction along the ballast water. Although there is no pressure loss and flow rate reduction, swirl filtration can filter out sediments such as sediments of 50 μm or more. However, it is difficult to filter out planktons such as mycelium and algae that are lighter than seawater, and ultrasonic cleaning is required to use ultrasound. The phytoplankton such as mycelium and algae were crushed and secondary filtration was performed. Therefore, the success or failure of these two filter designs directly affects the product performance of the ballast water and marine biological management system.

6.1.1 Backwash Filter Optimization

The optimization of the backwash filter is mainly to increase the flow area of the filter, and keep the filter in the cleaning process, so that the backwash filter will not deposit filter cakes during the working process, which, as a result, will not increasethe flow resistance in ballast pipes and will not reduce the flow.

The optimization of the backwash filter, from the perspective of increasing the filter area, is usually done by selectinga columnar filter cartridge, as shown in Fig. 6.1, or changing the cylindrical filter to several small cylindrical filters, as shown in Fig. 6.2. The only purpose of these two improved methods is to increase the flow area of the filter.

From the point of view of the two types of filter elements, due to seawater turbidity and the effect of static pressure, impurities will adhere to the filter mesh filter to form filter residue, requiring continuous cleaning to remove the filter residue in order

N. Chen et al., *Development and Implementation of Ship BWMS*,
https://doi.org/10.1007/978-981-33-6865-1_6

Fig. 6.1 A columnar element with fold

Fig. 6.2 Several small cylindrical filters

to ensure the permeability of the filter. Apparently, although the filter made of the filter screen increases its surface area, when the cleaning nozzle rotates and the filter deposition surface is cleaned up and down, the negative suction force formed by the suction head cannot effectively suck due to the far distance separating the suction head from the fold portion. The continuous operation of the ballast water system requires that its filter element cannot be replaced. Therefore, the use of a filter element with a pleated filter is obviously not suitable for the pre-filter design of the ballast water biological control system. At present, almost all of the ballast water anti-sea biological management system equipment manufacturers focus on the improvement of the backwash filter in the development of multi-filter filter.

As the column-type multi-filter backwash filter is in operation, the motor at the top of the backwash manager drives the sewage pipe to rotate, which in turn causes the cylindrical filter element to open to the atmosphere, reducing the internal pressure of the circuit, forming a siphon phenomenon, and discharging the filter residue attached to the filter. Ensuring the smooth flow of the filter elements ensures that the head and flow of ballast water are not lost, as shown in Fig. 6.3.

Multi-filter automatic backwash filters improve the filtration process of ballast water, but for ships engaged in ballast operations in environments with large

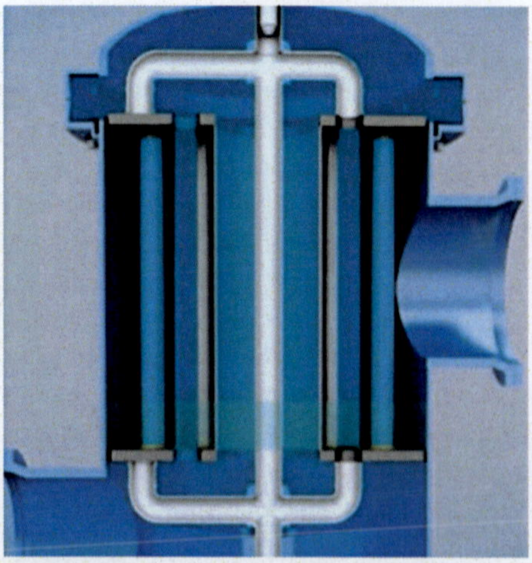

Fig. 6.3 Loop of discharge pipe and filter element

sediment-bearing waters, such filters do not fundamentally solve the problem of clogging of filters. For this reason, people have invented a sewage pipe fixing, and the filter top disk is driven by the top motor of the filter to form a sewage circuit with the suction pipe, and an ultrasonic cleaner or an external high-pressure water cleaning nozzle is installed on both sides of the filter symmetrically to the fixed sewage pipe, and the rotation is to this position. The mechanism for reverse cleaning of the filter element rotated to this position, as shown in Fig. 4.5a.

6.1.2 Swirl Filter Optimization

Swirl filter optimization is mainly through the change of its structural shape and size, to achieve the greatest possible removal of sediment and microorganisms in ballast water.

The optimization of the swirl filter is mainly accomplished through the numerical calculation of the solid–liquid two-phase flow in it.

Assume that the flow medium during the operation of the swirl filter is mainly a two-phase mixture of water and sediment. The FLUENT software was used to numerically simulate the cyclone, and a numerical calculation model of solid–liquid two-phase flow was established for the complex two-phase turbulent flow characteristics inside the swirl filter.

1. **Establishment of solid–liquid two-phase flow model**

FLUENT contains three kinds of multiphase flow models, namely VOF, Mixture model, and Eulerian model. Since the VOF model is only applicable to the simulation of layered or free surface flow and is not suitable for the complex two-phase flow simulation in the swirl filter, it will not be discussed here.

The Mixture model is a simplified multiphase flow model that can be used to simulate multi-phase flow with relative motion between phases. The model treats each phase as a single-phase flow, regardless of the interaction, and uses the concept of slip velocity to solve the mixed-phase momentum equation, continuity equation, and energy equation [1].

The Eulerian model is generally used to simulate the interaction of multiphase flow separations. The model establishes the local conservation equations for each phase, uses the average method to obtain the basic equations of multiphase flow, introduces the particle dynamics model, fully considers the force of the particles in the flow field, and solves the continuous equations of each phase separately. Momentum equations and turbulent control equations can reflect the actual situation of the flow field in a more comprehensive and reasonable way, and can well describe the turbulent mixing process of the solid phase and the fluid.

The three principles of multi-phase flow model selection [2] are:

(1) If the disperse phase is more concentrated, then select the Eluerian model, otherwise choose the Mixture model;
(2) Eluerian model is selected when the force of each phase is relatively regular, otherwise Mixture model is selected;
(3) Under normal circumstances, the calculation stability of the Mixture model is better, and the calculation of the Eluerian model has higher accuracy.

In summary, because the solid–liquid two-phase flow studied in this example is a typical flow separation phenomenon, and the dispersed phase is relatively concentrated in the swirl filter, so in order to more accurately simulate the distribution of two-phase flow inside the swirl filter, the multi-phase flow model in this example will use the Eluerian model.

The continuous equation of the hydro-cyclone solid–liquid two-phase flow model was established according to the Euler model. Here is the liquid continuous equation:

$$\frac{\partial}{\partial x_i}(\alpha_l \rho_l u_{l,i}) = 0 \tag{6.1}$$

Solid-phase continuous equation:

$$\frac{\partial}{\partial x_i}(\alpha_s \rho_s u_{s,i}) = 0 \tag{6.2}$$

where:

α_l, α_s Liquid and solid volume fractions;
$u_{l,i}$, $u_{s,i}$ Liquid phase, solid phase velocity, i is the tensor indicator, $i = 1, 2, 3$;
ρ_l, ρ_s Liquid and solid phase densities.

The solid phase (sediment) diameters in ballast water are different. In the simulation, sediments with different diameters can be regarded as different phases. Therefore, the flow in the cyclone can be regarded as the multiphase flow of a liquid phase and multiple solid phases. Consider the interaction forces and establish momentum equations for liquid and solid phases [3], as shown in the liquid momentum equation below:

$$\frac{\partial}{\partial x_j}(\alpha_l \rho_l u_{l,i} u_{l,j}) = -\frac{\partial}{\partial x_i}(\alpha_l p) + \frac{\partial \tau_{l,ij}}{\partial x_j} + \alpha_l \rho_l g_i + \sum_{p=1}^{n} \frac{\partial}{\partial x_j}[K_{pl}(u_{p,j} - u_{l,j})] + F_{lift,l,i} + F_{Vm,l,i} + F_{l,i} \quad (6.3)$$

Solid-state momentum equation:

$$\frac{\partial}{\partial t}(\alpha_s \rho_s u_{s,i}) + \frac{\partial}{\partial x_j}(\alpha_s \rho_s u_{s,i} u_{s,j}) = -\frac{\partial}{\partial x_i}(\alpha_s p) - \frac{\partial}{\partial x_i}(p_s) + \frac{\partial \tau_{s,ij}}{\partial x_j} + \alpha_s \rho_s g_i + \sum_{q=1}^{n} \frac{\partial}{\partial x_j}[K_{qs}(u_{q,j} - u_{s,j})] + F_{lift,s,i} + F_{Vm,s,i} + F_{s,i} \quad (6.4)$$

where:

p	the pressure shared by the two parties;
u_{li}, u_{si}	Liquid phase, solid phase velocity;
u_{pi}, u_{qi}	Phase p, phase q velocity;
g_i	Gravity acceleration;
p_s	Solid pressure;
$\tau_{l,ij}, \tau_{s,ij}$	Pressure and strain tensors in liquid and solid phases;
K_{pl}, K_{qs}	Momentum exchange coefficient between phases, K_{pl} For the first relative p liquid phase l momentum exchange coefficient, K_{qs} Momentum exchange coefficient between the q phase and the solid phase s, and $K_{pl} = K_{lp}, K_{qs} = K_{sq}$;
$F_{l,i}, F_{s,i}$	Gravity in the solid and liquid phases. When the cyclone is simulated, the centrifugal force is not greater than the weight of the swirl filter.
$F_{lift,l,i}, F_{lift,s,i}$	Liquid phase, solid phase lift, $F_{lift,l,i} = -F_{lift,s,i}$. Since the model in FLUENT assumes that the diameter of the particle is much smaller than the distance between the particles, lift is not considered;
$F_{Vm,l,i}, F_{Vm,s,i}$	Virtual mass force of liquid phase and solid phase, and $F_{Vm,l,i} = -F_{Vm,s,i}$. Inside the swirl filter, the acceleration of the particles is very small and this term is negligible when calculating.

2. **Solid-Liquid Two-Phase Turbulence Model**

The flow field in the swirl filter is a kind of high Reynolds number, strong swirling turbulent flow, and the streamline in the strong swirling field is characterized by high-speed rotation and rapid change. For turbulent flow, the Reynolds average method is often used to process the turbulent control equations. According to the different assumptions and treatment methods for Reynolds stress, turbulence models can be divided into standard k−ε, RNG k−ε and other eddy viscosity models and RSM models. The eddy viscosity model makes isotropic assumptions for turbulence and as a result, it cannot describe accurately the anisotropic swirling flow field [4]. The RSM model discards the assumption of isotropicity and strictly considers the anisotropy of turbulent viscosity and various factors that may cause Reynolds stress to change, such as turbulent kinetic energy, turbulent kinetic energy dissipation rate, rapid changes in tension, rotationetc. Pressure tensor and dissipation rate are considered as the main factors to reduce the accuracy of flow field prediction [5]. Therefore, the RSM model has more accurate prediction and description capabilities for complex flow fields and is more suitable for the strong swirling flow problem in this book. Based on above considerations, the turbulence model in this bookuses the RSM model.

The differential equation of turbulent pulsating stress in the Reynolds equation is established and solved. The instantaneous control equation is processed using the Reynolds time-averaged method, assuming that the turbulent flow is a superposition of the time-averaged flow and the instantaneous pulsating flow [6]. The superscript "-" represents the mean value, and the superscript "'" represents the pulsation value. The relationship between the hourly value $\overline{\phi}$ and the pulsation value ϕ' is as follows:

$$\phi = \overline{\phi} + \phi' \tag{6.5}$$

According to the time-average method, the flow variable is expressed as the sum of the mean value and the pulsation value. That is:

$$u = \overline{u} + u';\ v = \overline{v} + v';\ w = \overline{w} + w';\ p = \overline{p} + p' \tag{6.6}$$

Applying Reynolds time-averaging to Eqs. (6.1)–(6.4), regardless of fluid, particle density pulsation, fluid velocity pulsation and particle mass pulsation, and removing the mean sign,the following equations can be obtained:

Liquid continuous equation:

$$\frac{\partial}{\partial x_i}(\alpha_l \rho_l u_{l,i}) = 0 \tag{6.7}$$

Solid-phase continuous equation:

$$\frac{\partial}{\partial x_i}(\alpha_s \rho_s u_{s,i}) = -\frac{\partial}{\partial x_i}\left(\overline{\rho_s' u_{s,i}'}\right) \tag{6.8}$$

Liquid momentum equation:

$$\frac{\partial}{\partial x_j}(\alpha_l \rho_l u_{l,i} u_{l,j}) = -\frac{\partial}{\partial x_i}(\alpha_l p) - \frac{\partial}{\partial x_j}\left(\overline{u'_{l,i} u'_{l,j}}\right) + \alpha_l \rho_l g_i + \frac{\partial}{\partial x_j}\left[K_{ls}\left(u_{l,j} - u_{s,j}\right)\right] \tag{6.9}$$

Solid-state continuous equation:

$$\frac{\partial}{\partial x_j}(\alpha_s \rho_s u_{s,i} u_{s,j}) = -\frac{\partial}{\partial x_i}(\alpha_s p) - \alpha_s K_{sl}(u_{l,i} - u_{s,i}) + \alpha_s \rho_s g_i + K_{sl}\left[(u_{s,j} - u_{l,j}) - \frac{1}{\rho_s}\left(\overline{\rho'_s u'_{s,i}}\right)\right]$$
$$- \frac{\partial}{\partial x_j}\left(\rho_s \overline{u'_{s,j} u'_{s,i}} + u_{s,i}\overline{\rho'_s u'_{s,j}} + u_{s,j}\overline{\rho'_s u'_{s,i}}\right) \tag{6.10}$$

In the above equations, $\overline{\rho'_s u'_{s,i}}$ and $\overline{\rho'_s u'_{s,j}}$ are the solid-density pulse fluxes, $\overline{u'_{l,i} u'_{l,j}}$ and $\overline{u'_{s,i} u'_{s,j}}$ are the liquid Reynolds stress and the solid Reynolds stress, respectively.With the introduction ofthe concept of scalar viscosity coefficient, the following can be obtained:

$$\overline{\rho'_s u'_{s,i}} = \frac{v_s}{\sigma_s}\frac{\partial \rho_s}{\partial x_i} \tag{6.11}$$

$$\overline{\rho'_s u'_{s,j}} = \frac{v_s}{\sigma_s}\frac{\partial \rho_s}{\partial x_j} \tag{6.12}$$

It can be seen that the unknowns in the system of equations are only the Reynolds stress of the solid and liquid two phases. The Reynolds stress generated by the time-averaging can be calculated using the RSM hybrid turbulence model based on the Eluerian multiphase flow model. The turbulence model assumes that the turbulent pulsation flow field of each phase is the same, that is:

$$\overline{u'_i u'_j} = \overline{u'_{l,i} u'_{l,j}} = \overline{u'_{s,i} u'_{s,j}} \tag{6.13}$$

The mixing density is:

$$\rho_m = \alpha_l \rho_l + \alpha_s \rho_s \tag{6.14}$$

The mixing speed is:

$$u_m = \frac{\alpha_l \rho_l u_l + \alpha_s \rho_s u_s}{\alpha_l \rho_l + \alpha_s \rho_s} \tag{6.15}$$

In summary, the RSM two-phase turbulence model can be expressed as:

$$\frac{\partial}{\partial t}\left(\rho_m\overline{u_i'u_j'}\right)+\underbrace{\frac{\partial}{\partial x_k}\left(\rho_m u_{m,k}\overline{u_i'u_j'}\right)}_{C_{ij}}=\underbrace{-\rho_m\left(\overline{u_i'u_k'}\frac{\partial u_{m,j}}{\partial x_k}+\overline{u_j'u_k'}\frac{\partial u_{m,i}}{\partial x_k}\right)}_{P_{ij}}+\underbrace{\frac{\partial}{\partial x_k}\left[\mu_m\frac{\partial}{\partial x_k}\left(\overline{u_i'u_j'}\right)\right]}_{D_{L,ij}}$$
$$-\underbrace{\frac{\partial}{\partial x_k}\left[\frac{\partial}{\partial x_k}\left(\rho_m\overline{u_i'u_j'u_k'}\right)\right]}_{D_{T,ij}}+\underbrace{\overline{P\left(\frac{\partial u_i'}{\partial x_j}+\frac{\partial u_j'}{\partial x_i}\right)}}_{\Phi_{ij}}-\underbrace{2\mu_m\overline{\frac{\partial u_i'}{\partial x_k}\frac{\partial u_j'}{\partial x_k}}}_{\varepsilon_{ij}} \tag{6.16}$$

where: the first item is an instantaneous item; C_{ij} Convection item; $D_{T,ij}$ turbulent diffusion items, $D_{T,ij}=\frac{\partial}{\partial x_k}\left(\frac{\mu_t}{\sigma_k}\frac{\partial\overline{u_i'u_j'}}{\partial x_k}\right)$, μ_t turbulent viscosity, $\mu_t=\rho_m C_\mu\frac{k^2}{\varepsilon}$, $k=\frac{1}{2}\overline{u_i'u_j'}$, $\sigma_k=0.82$, $C_\mu=0.09$; $D_{L,ij}$ Molecular adhesion diffusion term; P_{ij} Shear stress generated items; Φ_{ij} Stress and strain items; ε_{ij} For viscous dissipative items, $\varepsilon_{ij}=\frac{2}{3}\rho_m\varepsilon\delta_{ij}$.

In the Reynolds stress equation, the turbulent kinetic energy k and the dissipation rate ε are included. The equation is as follows:

Kinetic energy:

$$\frac{\partial(\rho_m k)}{\partial t}+\frac{\partial(\rho_m k u_{m,i})}{\partial x_i}=\frac{\partial}{\partial_j}\left[\left(\mu_m+\frac{\mu_t}{\sigma_k}\right)\frac{\partial k}{\partial x_j}\right]+\frac{1}{2}P_{ij}-\rho_m\varepsilon \tag{6.17}$$

Dissipation rate:

$$\frac{\partial(\rho_m\varepsilon)}{\partial t}+\frac{\partial(\rho_m\varepsilon u_{m,i})}{\partial x_i}=\frac{\partial}{\partial_j}\left[\left(\mu_m+\frac{\mu_t}{\sigma_\varepsilon}\right)\frac{\partial k}{\partial x_j}\right]+C_{1\varepsilon}\frac{1}{2}P_{ij}-C_{2\varepsilon}\rho_m\varepsilon \tag{6.18}$$

where: $C_{1\varepsilon}=1.44$, $C_{2\varepsilon}=1.92$, $\sigma_\varepsilon=1$.

In summary, Eqs. (6.1)–(6.4) and (6.15)–(6.18) constitute the hydro-cyclone solid–liquid two-phase turbulence calculation model.

3. Geometric model construction and meshing

The geometric model of the hydrocyclone is appropriately simplified. According to the structural dimensions in Table 5.5, the hydrocyclone fluid domain model is established, as shown in Fig. 6.4a. Because the RSM model requires very strict grid division, in order to obtain a high-quality grid, this book adopts ICEMCFD to divide the model into sub-area grids, and generate a three-dimensional structured grid as shown in Fig. 6.4b.

4. Boundary conditions and setting of control parameters

(1) Liquid boundary conditions

① **Inlet boundary**: The entrance boundary is defined as the velocity entrance, taking the average velocity perpendicular to the entrance face:

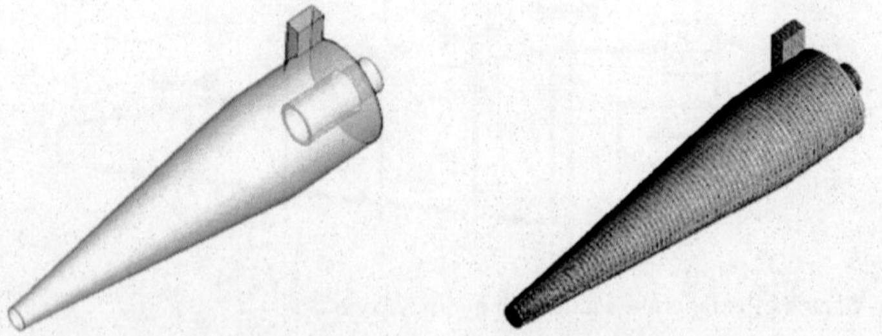

Fig. 6.4 The geometric model and mesh diagram of Hydrocyclone

$$u_l = \frac{q_m}{A_{in}} \tag{6.19}$$

After some calculation, one gets $u_l = 6.57$ m/s. Turbulence intensity $I = 0.16(\mathrm{Re}_{D_H})^{-1/8}$, where D_H is the hydraulic diameter, and the rectangular hydraulic diameter is 4 times the sectional area divided by the wet period.

② **Outlet boundary**:

Overflow port and underflow port are respectively set as pressure outlets, and they are all open to the atmosphere. The pressure is set to 0.

③ **Wall conditions**:

Non-slip solid wall conditions.

(2) **Solid boundary conditions**

① **Inlet boundary**: set the entrance boundary to the speed entrance, speed $u_s = u_l$; Inlet solid phase volume fraction α_s takes 5%.

② **Outlet boundary**: Overflow port and underflow port are respectively set as pressure outlets, and they are all open to the atmosphere. The pressure is set to 0.

③ **Wall condition**: Non-slip solid condition.

(3) **Control parameters**

The water was set as the basic phase, the density was 998.2 kg/m^3, the viscosity was 0.001003 Pa·s, the sediment was solid phase, the density was 2650 kg/m^3, the particle size was 60 μm, and the mass flow rate was 8.327 kg/s. For the diffusion term, the central difference scheme and convection phase are used. Second-order upwind and QUICK difference schemes are used to separate the control equations. The SIMPLEC algorithm is used to solve the velocity–pressure field. The pressure interpolation format is PRESTO. Firstly, the RSM model is used to calculate the steady flow of

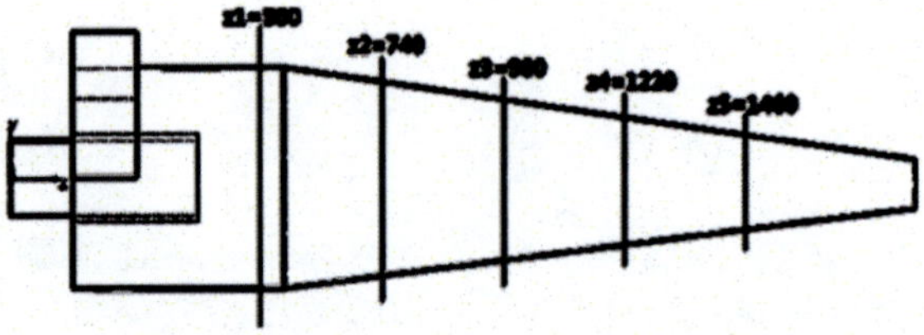

Fig. 6.5 Sketch of axial cross-sections in hydrocyclone

the basic phase. After the convergence of the results, the Euler model is activated and the numerical simulation of the solid–liquid two-phase flow is completed.

4. **Flow field distribution in the swirl filter**

In order to better observe and analyze the velocity distribution, pressure distribution and solid volume fraction distribution in the swirl filter, five cross sections were taken in the column section and the cone section respectively, i.e. z1 = 500 mm, z2 = 740 mm, z3 = 980 mm., z4 = 1220 mm, z5 = 1460 mm.The origin of the coordinate is located in the overflow mouth section, where z = 0. The cross-sectional distribution is shown in Fig. 6.5.

(1) **Pressure distribution.**

As shown in Fig. 6.6, the static pressure is distributed symmetrically across the cross-axis. The overall distribution trend is that from the wall to the axis, the static pressure gradually decreases, and the isobar is approximately parallel to the axis. From the cyclone column section to the cone section, the static pressure gradually decreases. This is due to the fact that the static pressure during the movement can be converted into kinetic energy to maintain the fluid movement, which promotes the flow of fluid along the wall toward the bottom flow, and the fluid and wall Friction losses also cause the constant reduction of static pressure.

The pressure distribution of each section is shown in Fig. 6.7. It can be seen that under the radial transmission of centrifugal force, a radial pressure difference occurs in the fluid, and the static pressure decreases monotonously along the wall toward the central axis. In the inner swirl area of the center, the static pressure loss is intense

Fig. 6.6 Static pressure distribution nephogram

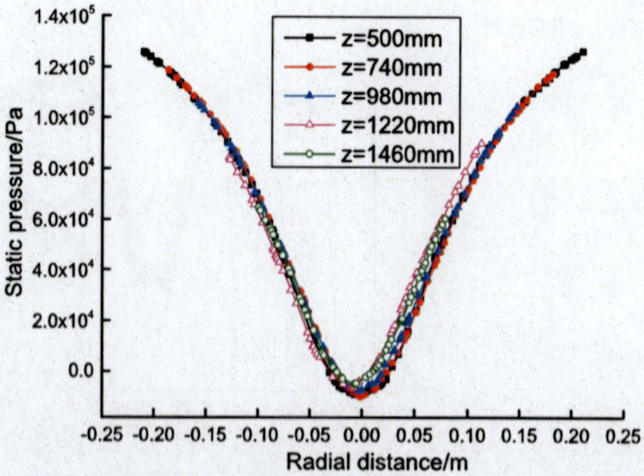

Fig. 6.7 Static pressure distribution of different sections

and negative pressure occurs. In the axial direction, the pressures on the same radius are nearly equal, which indicates that the axial pressure gradient is almost zero.

(2) **Speed field distribution.**

① **Tangential speed.**

The tangential velocity provides the required centrifugal force for the solid and liquid separation in the cyclone, and its value determines the separation performance of the swirl filter [7]. Figures 6.8 and 6.9 are the tangential velocity distribution within the swirl filter and the tangential velocity distribution along each section. As can be seen from Fig. 6.8, the tangential velocity is axisymmetric, and near the overflow diameter, the region of maximum tangential velocity appears. Within the maximum speed, it is the inner swirl area, and the tangential speed increases with the increase of radius. Outside the maximum speed, it is the outer swirl area, and the tangential speed increases with the reduction of radius.

As can be seen in Fig. 6.9, stable strong swirling flow has been formed at the z = 500 mm section of the column segment, which indicates that the designed cyclone can provide the required tangential velocity for the fluid. Due to the presence of a boundary layer at the wall of the cyclone, the tangential velocity on the wall varies

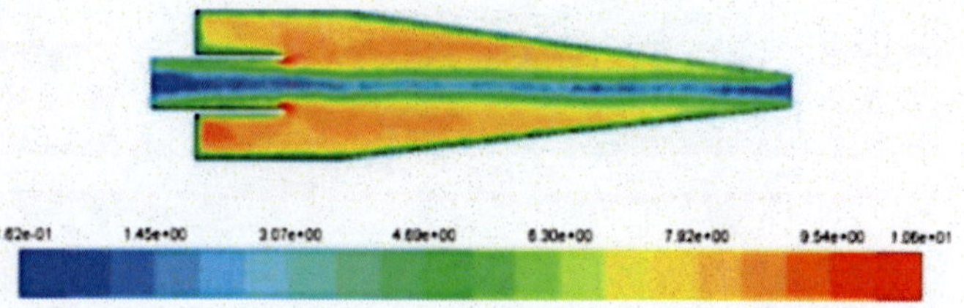

Fig. 6.8 Tangential velocity distribution

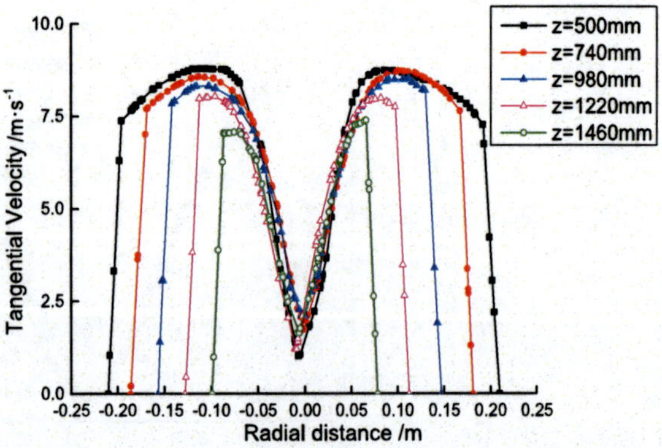

Fig. 6.9 Tangential velocity distribution

drastically. In addition, except for the z = 1460 mm section, the maximum tangential velocity on the different sections in the swirl cone is not much different. This shows that the swirl cone has better acceleration effect and small velocity loss, which is beneficial to the separation.

② **Axial speed.**

Relative to the tangential velocity, the numerical value of the axial velocity is smaller and more complex, and the flow direction determines the distribution of the fluid in the overflow and underflow. The swirl filter's axial velocity distribution cloud is shown in Fig. 6.10. It can be seen from above that the velocity of the fluid in the near wall region is positive and downward, and the velocity of the fluid in the central region is negative and upward.

Figure 6.11 shows the axial velocity distribution on each section. It can be seen from the figure that the axial velocity is symmetrical and the velocity changes from positive to negative in the direction from the wall to the central axis, and the velocity value gradually increases, and in the middle of the radius it crossesthezero point, at which the axial velocity is zero, which can be plotted as a conical surface, the zero axial velocity envelope (LZVV) [8]. The zero-axial envelope surface within the cyclone can be divided into two swirl zones: the fluid flow in the inner swirl zone flows to the overflow at the edge, and is finally discharged from the overflow port; the velocity of the fluid in the the outer swirl zone is downwards, and the fluid

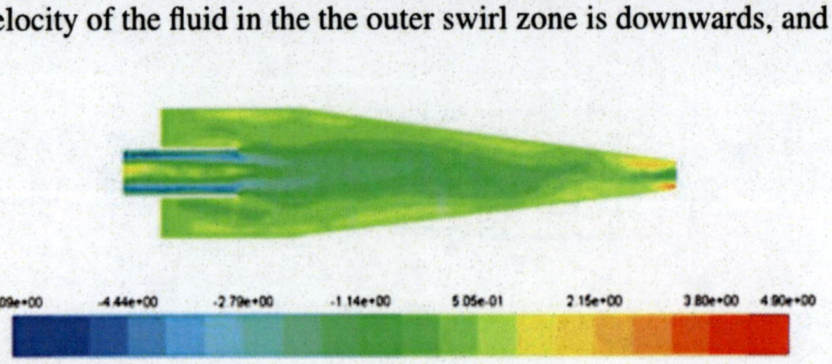

Fig. 6.10 Axial velocity distribution

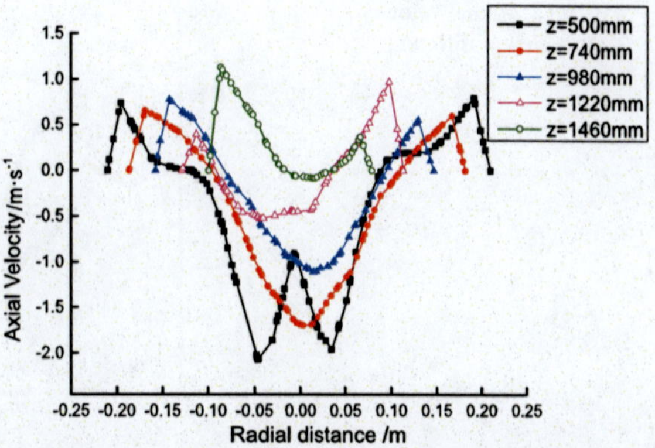

Fig. 6.11 Axial velocity distribution of different sections

flows while swirling toward the underflow port, and is eventually discharged from the underflow. As the section moves downwards, the positive axial velocity of the outer swirling flow increases gradually, and the axial velocity on the section of the cone section shows asymmetry. This is because the axial velocity is greatly affected by turbulence and eddy currents in the flow field.

③ **Radial speed**.

In the three-dimensional flow of the swirl filter, the radial velocity is numerically much smaller than the other two-direction velocity. The radial velocity reflects the movement law of the fluid in the radial direction. It is the main factor of the resistance of the separation phase settlement, and has an important influence on the separation effect of the swirl filter. The radial velocity distribution in the cyclone is shown in Figs. 6.12 and 6.13.

As can be seen from the figures above, the radial velocity of the fluid in the column region of the swirl filter fluctuates greatly. From the outer wall to the overflow, the radial velocity is positive, indicating that the fluid moves in a direction away from the axis, favoring the solid phase. The particles move outwards into the outer swirl; from the overflow to the center axis, the radial velocity becomes negative, indicating that the fluid moves in the direction of the axis. After the fluid reaches the cone, the regularity of radial velocity distribution increases, indicating that the fluid movement at this time has stabilized; from the wall of the device to the central axis, the radial

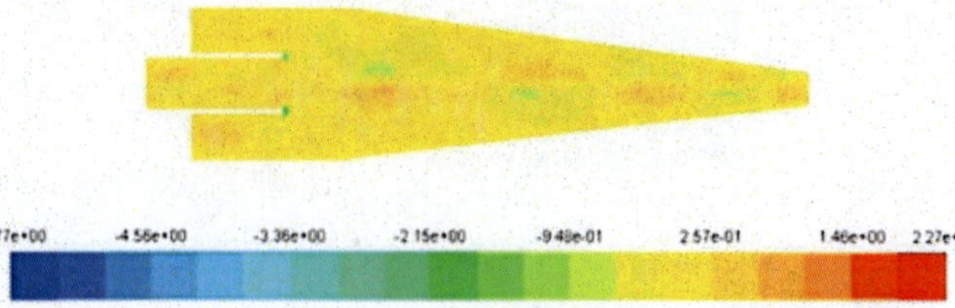

Fig. 6.12 Radial velocity distribution

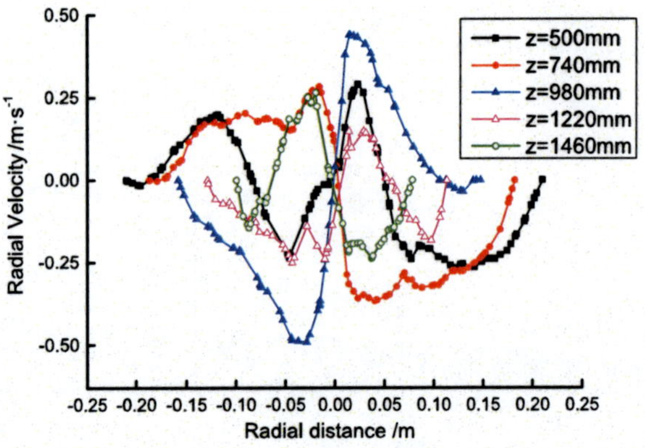

Fig. 6.13 Radial velocity distribution of different sections

velocity is negative due to the effect of the cone, and it moves toward the center; As the radius decreases, the radial velocity decreases first and then increases, and becomes positive near the center of the swirl flow.

(2) **Solid phase concentration distribution**

Figures 6.14 and 6.15 are the volume fraction distribution cloud diagrams of the solid phase inside the swirl filter and the solid volume fraction distribution maps of each section. As can be seen from Fig. 6.14, when the solid–liquid mixture enters the filter from the inlet, the solid phase separates under centrifugal force, and the solid phase is mainly distributed in the outer wall region after separation. Therefore, in the filter, the solid phase concentration near the wall region is higher, and the solid phase concentration near the axis region is lower. It can be seen from Fig. 6.15 that as the cross section moves down, the solid phase concentration gradually increases, and the solid phase concentration in the cone section is greater than the column section. This shows that the two-phase separation in the filter is mainly performed in the cone

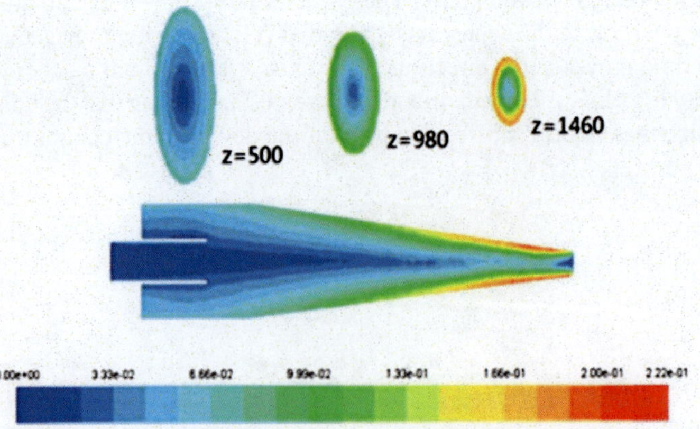

Fig. 6.14 Solid volume fraction distribution

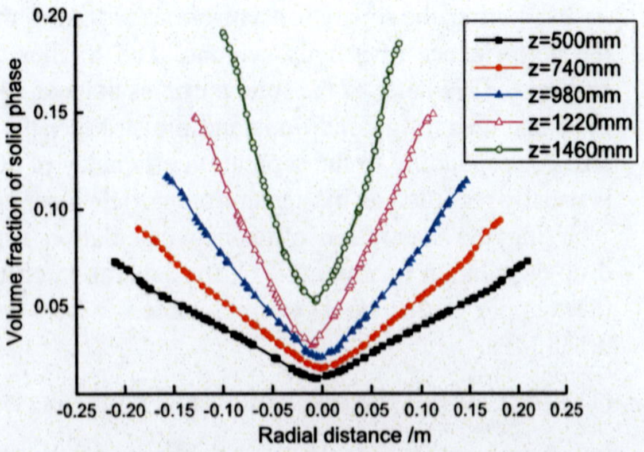

Fig. 6.15 Solid volume fraction distribution of each section

section. The distribution law of the volume fraction of solid phase fully shows that the filter designed in this book can realize the effective separation of solid and liquid.

5. **Influence of structural parameters' change on separation performance and optimization of structural parameters**

Separation efficiency and pressure drop are important indicators for evaluating the performance of a swirl filter.

For the swirl filter, the separation efficiency is one of the important indicators to evaluate its performance, it is the main factor describing the separation effect [9]. The solid particles separated by the cyclone are discharged from their underflow ports. Therefore, the separation efficiency can be defined as the percent removal of solid particles at the underflow port. The separation efficiency of a certain diameter particle can be calculated as follows:

$$E = \frac{M_u}{M_i} \times 100\% \tag{6.20}$$

where:

M_u Mass flow rate of solid particles at the underflow port (kg/s);
M_i Mass flow rate of solid particles at the inlet (kg/s);
E Separation efficiency (%).

There is no power device in the filter itself. The flow and separation process of the fluid in the cyclone must rely on the pressure loss of the internal fluid to obtain power. Therefore, the pressure drop is another important indicator to measure the energy consumption of the cyclone. With the same processing capacity and separation efficiency, the pressure drop should be as low as possible.Let the inlet pressure be P_i, Overflow pressure is P_o, Overflow pressure drop ΔP is:

$$\Delta P = P_i - P_o \tag{6.21}$$

In addition, the separation particle size is one of the main indicators for evaluating the performance of a hydrocyclone. The smaller the separation particle size, the smaller the diameter of the solid particles that can be separated by the cyclone under the same operating conditions and the stronger the separation ability. The particle size corresponding to the separation efficiency of 50% at the stage efficiency curve is usually recorded as the separation particle size, which is represented by d_{50c} [9].

Using the separation efficiency calculation formula (6.20) and the pressure drop calculation formula (6.21), the separation efficiency of the initial structure of theexample in this book was calculated to be 66.6% and the pressure drop was 0.138 MPa.

(1) Effect of wall thickness of overflow pipe on separation performance

In general, increasing the wall thickness of the overflow pipe appropriately can improve the separation efficiency of the swirl filter. In this example, five kinds of cyclones with different overflow pipe wall thicknesses were analyzed. The overflow pipe wall thickness was set to 7.5 mm, 10 mm, 12.5 mm, 15 mm, and 17.5 mm, respectively. The other structural parameters were the same as the initial structure. Under the same conditions, numerical simulations were performed separately. Figures 6.16 and 6.17 show the effect of overflow wall thickness on separation efficiency and pressure drop. It can be seen from the figures that the separation efficiency increases first with the increase of wall thickness of the overflow pipe. When the wall thickness increases to a certain value, the separation efficiency begins to decrease with the increase of the wall thickness of the overflow pipe, and when the wall thickness increases to 15 mm, the separation efficiency is the highest, which is 66.6%. The pressure drop increases first and then decreases with the increase of thewall thickness of the overflow pipe. When the wall thickness is 15 mm, the pressure drop is 0.138 MPa.

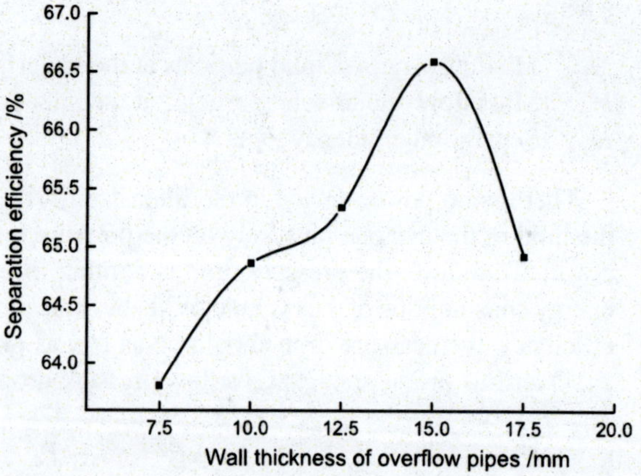

Fig. 6.16 Effect of wall thickness on separation efficiency

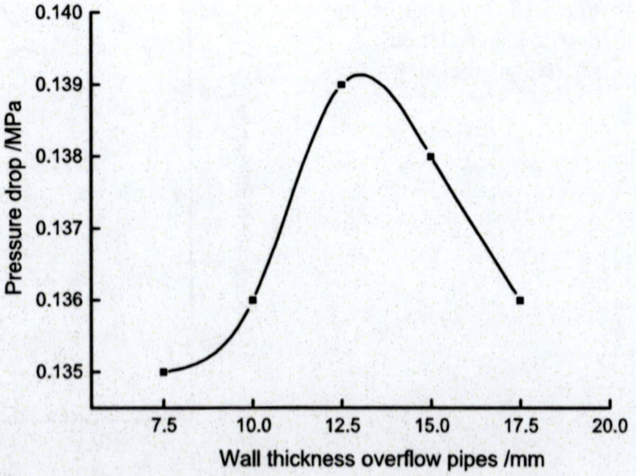

Fig. 6.17 Influence of overflow pipe wall thickness on pressure drop

(2) **Effect of Overflow Pipe Diameter on Separation Performance**

The size of the overflow diameter determines the distribution of the fluid between the underflow and the overflow, thereby affecting the separation performance. As the overflow diameter increases, the pressure of the overflow decreases, the fluid tends to flow out of the overflow, and the short-circuit flow is also more likely to occur. As a result, more fine-grained solid particles will flow from the overflow with the short-circuit flow, reducing the separation efficiency; As the diameter of the overflow decreases, the turbulence intensity near the overflow port increases, and the circulation flow increases, resulting in a decrease in separation efficiency. Therefore, for a given particle size, there will be an optimal overflow diameter. The wall thickness of the overflow pipe is set to 15 mm. The other structural parameters are the same as those of the initial structure. The overflow diameters are 120 mm, 130 mm, 140 mm, 150 mm and 160 mm, respectively, and they are calculated under the same conditions. Figures 6.18 and 6.19 show the separation efficiency and pressure drop curves for different overflow diameters. The analysis shows that with the increase of the overflow diameter, the separation efficiency increases first and then decreases, and the pressure drop decreases. When the overflow diameter is 130 mm, the maximum separation efficiency is 77.54% and the pressure drop is 0.149 MPa.

(3) **Effect of Column Length on Separation Performance**

Swirl filter column sections primarily serve to pre-swirl, stabilize, and buffer fluids. When the column length is small, the fluid enters the conical section in a more disordered state, which increases the turbulence intensity and also enhances the effects of the short-circuit flow and the circulation flow, thereby reducing the separation efficiency. When the column length is long, it will lead to the decrease in the tangential velocity of the fluid, which is not conducive to the separation, so the column length should be chosen to be the appropriate length. The overflow pipe diameter and wall thickness were set to 130 mm and 15 mm, respectively, and the column lengths were taken as 330 mm, 360 mm, 390 mm, 420 mm, and 460 mm, respectively. The

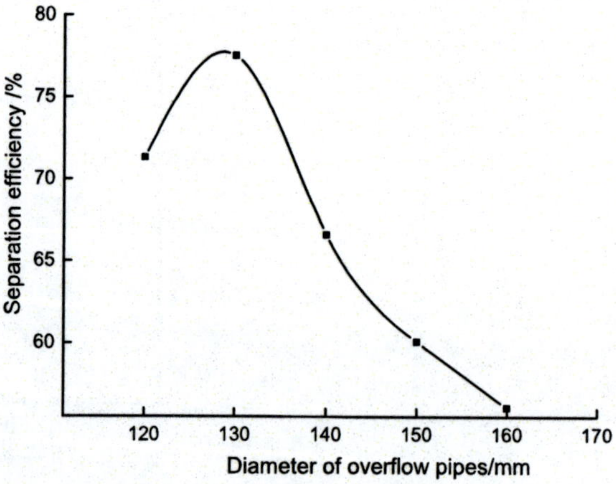

Fig. 6.18 Influence of overflow diameter on separation efficiency

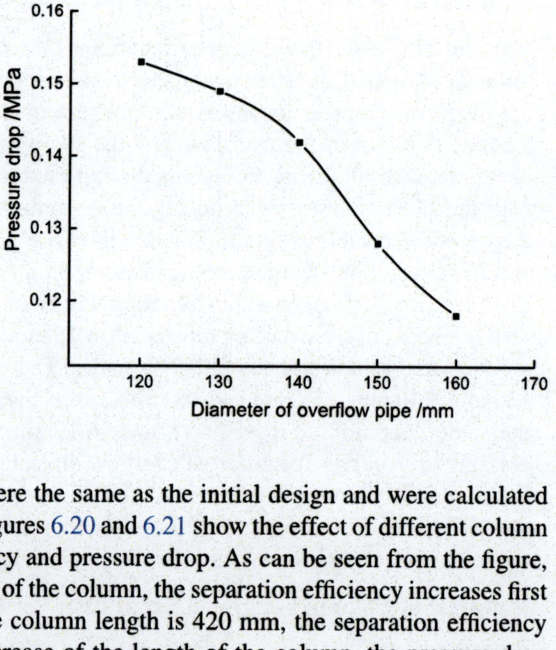

Fig. 6.19 Influence of overflow orifice diameter on pressure drop

other structural parameters were the same as the initial design and were calculated under the same conditions. Figures 6.20 and 6.21 show the effect of different column lengths on separation efficiency and pressure drop. As can be seen from the figure, with the increase of the length of the column, the separation efficiency increases first and then decreases. When the column length is 420 mm, the separation efficiency is up to 77.54%. With the increase of the length of the column, the pressure drop tends to decrease, but the change is relatively small, and when the column length is 420 mm, the pressure drop is 0.149 MPa.

(4) **Influence of cone angle on separation performance**

The proper reduction of the cone angle can increase the axial velocity and the tangential velocity of the inner and outer swirling flow in the cone, and promote the solid phase separation of the small diameter. However, if the cone angle is too small, the vortex intensity in the cone section will increase and the maximum tangential

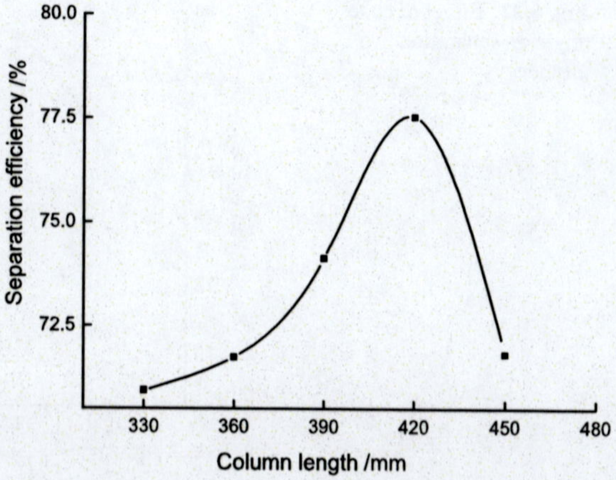

Fig. 6.20 Effect of column length on separation efficiency

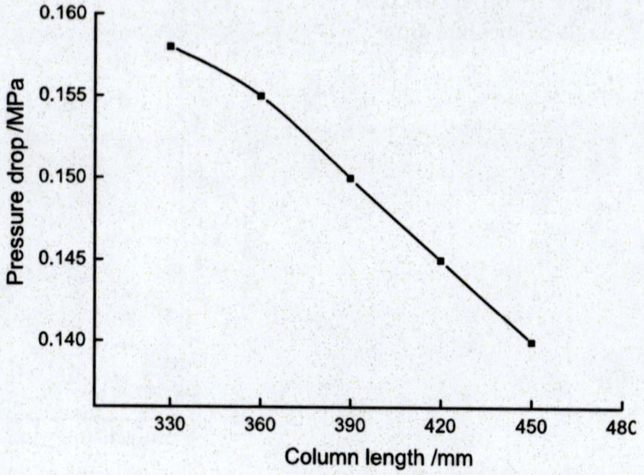

Fig. 6.21 Effect of column length on pressure drop

velocity will decrease, which is not conducive to the separation of the solid phase. So for a particular cyclone, there will be an optimum cone angle. The wall thickness of the overflow pipe is chosen to be 15 mm, the diameter of the overflow pipe is 130 mm, the length of the column segment is 420 mm, and the cone angles are set to 10, 12.5, 15, 17.5, and 20°. Other structural parameters are the same as the initial design. Under the same conditions, numerical simulations were performed. Figures 6.22 and 6.23 give the separation efficiency and pressure drop curves for different cone angles. As can be seen from the figures, with the increase of the cone angle, the separation efficiency increases first and then decreases. When the cone angle is 12.5°, the separation efficiency is up to 79.24%. With the increase of the cone angle, the pressure drop will increase to some extent. When the cone angle is 12.5°, the pressure drop is about 0.13 MPa.

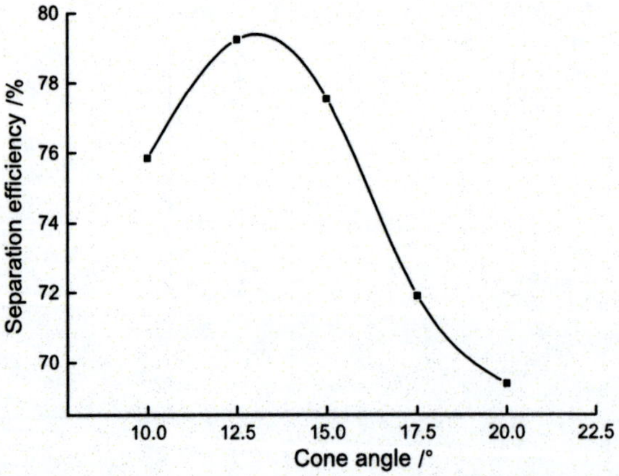

Fig. 6.22 Effect of cone angle on separation efficiency

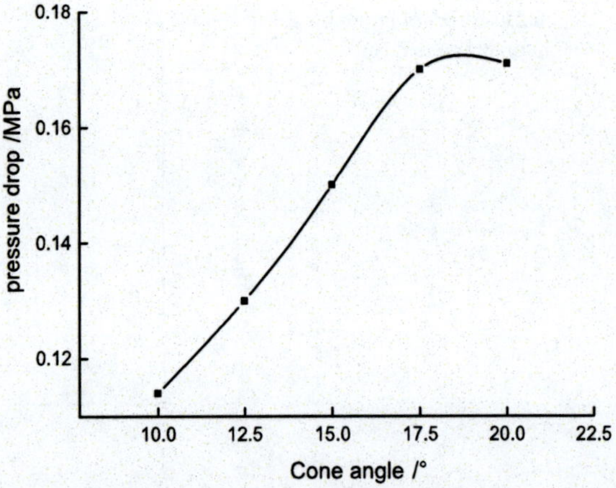

Fig. 6.23 Effect of cone angle on pressure drop

(5) **Effect of underflow opening diameter on separation performance**

The increase of the diameter of the underflow port will cause the pressure in the flow field to decrease, which can reduce the coarse particles in the overflow product to a certain extent and improve the separation efficiency of the coarse particles. However, if the diameter of the underflow port is too large, the separation effect will worsen. If the diameter is too small, the tangential velocity will decrease, which is not conducive to separation. Therefore, for a particular cyclone, there is a suitable underflow orifice diameter. The wall thickness of the overflow pipe is 15 mm, the diameter of the overflow pipe is 130 mm, the length of the column segment is 420 mm, the taper angle is 12.5°, and the underflow port diameter is set to 85 mm, 90 mm, 95 mm, 100 mm and 105 mm, respectively. Other structural parameters are the same as the initial design, and numerical simulation calculations are performed under the same conditions. Figure 6.24 and 6.25 show the separation efficiency and pressure drop

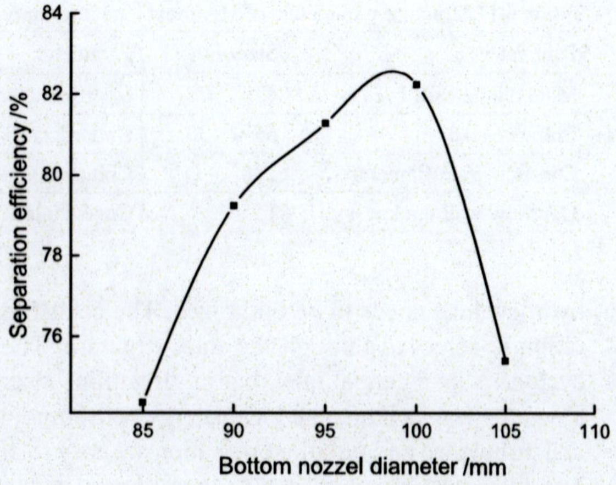

Fig. 6.24 Influence of underflow diameter on separation efficiency

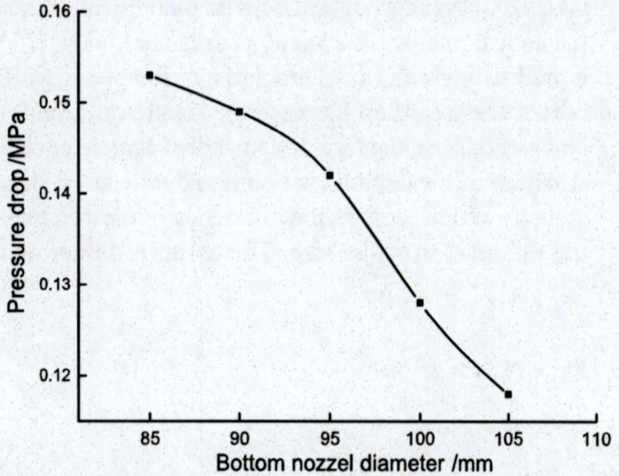

Fig. 6.25 Influence of apex diameter on the pressure drop

curves at different underflow orifice diameters. As can be seen from the figure, with the increase of the diameter of the underflow port, the separation efficiency increases first and then decreases. When the underflow port is 100 mm, the separation efficiency is the highest, which can reach 82.88%; with the increase of the taper angle, the pressure drop will occur to some extent, nonlinearly. Through the above analysis, the diameter of the underflow port should be set to 100 mm, and the pressure drop at this time is 0.128 MPa.

The preferred dimensions of the cyclone structure are shown in Table 6.1.

6. **Swirl inlet structure optimization**

The Ballast Water Management Convention D-2 clearly stipulates the scope and content of the particle matter in ballast water. To better meet this requirement, the

Table. 6.1 Structure parameters of Hydrocyclone after optimizing

Parameter	Size/mm	Parameter	Size/mm
Main diameter of cyclone	420	Insertion depth of overflow pipe	250
Entrance size	65 × 130	Underflow diameter	100
Overflow pipe diameter	130	Column length	420
Overflow wall thickness	15	Cone angle	12.5°

hydrocyclone needs to be optimized. The centrifugal force required for filter separation is achieved through the inlet structure. The common inlet structure of the cyclone is a tangential inlet, but its streamline changes abruptly, the resistance loss along the way is large, the local energy consumption caused by the cyclone is large, and turbulence and turbulent pulsation are easy to affect the separation process [10]. The filter inlet structure is also logarithmic, involute, concentric, and Archimedes spiral (hereinafter referred to as spiral). Gong Weian compared the resistance coefficients of these inlet structures with theoretical calculation methods. It is believed that the helical inlet has the smallest drag coefficient [11]. In addition, the spiral inlet has a gradually shrinking characteristic compared with the tangential inlet, which can reduce the wear and increase the rotational speed of the fluid and reduce the local energy consumption and the turbulent kinetic energy. Therefore, the tangential inlet structure of the cyclone is optimized as a spiral shape. In order to compare the flow velocity and the separation efficiency in the two inlet form cyclones, the two models use the same structure size. The optimized filter model is shown in Fig. 6.26.

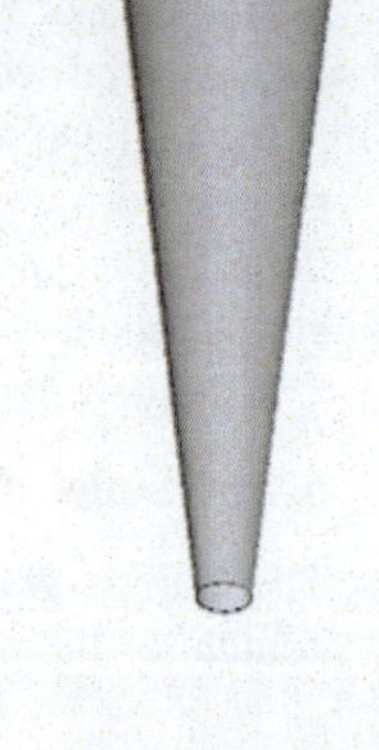

Fig. 6.26 The geometric model of Archimedes spiral Hydrocyclone

(1) Comparative analysis of tangential velocity

The tangential velocities of the two inlet structure swirlers on sections z1, z2, z3, z4 are compared, as shown in Fig. 6.27. It can be seen from the figure that the tangential velocity of the helical inlet is greater than that of the tangential inlet. The mean value of the tangential velocity of the spiral inlet on the z1 section is increased by 59.66% compared with the tangential inlet, and at the same time, The tangential speeds on thez2, z3, and z4 sections also increased by 58.45%, 41.73%, and 44.42%, respectively. The increase in the tangential speed means an increase in centrifugal force, which contributes greatly to the solid–liquid separation in the cyclone. Therefore, from the perspective of tangential velocity, the spiral inlet structure is superior to the tangential inlet structure.

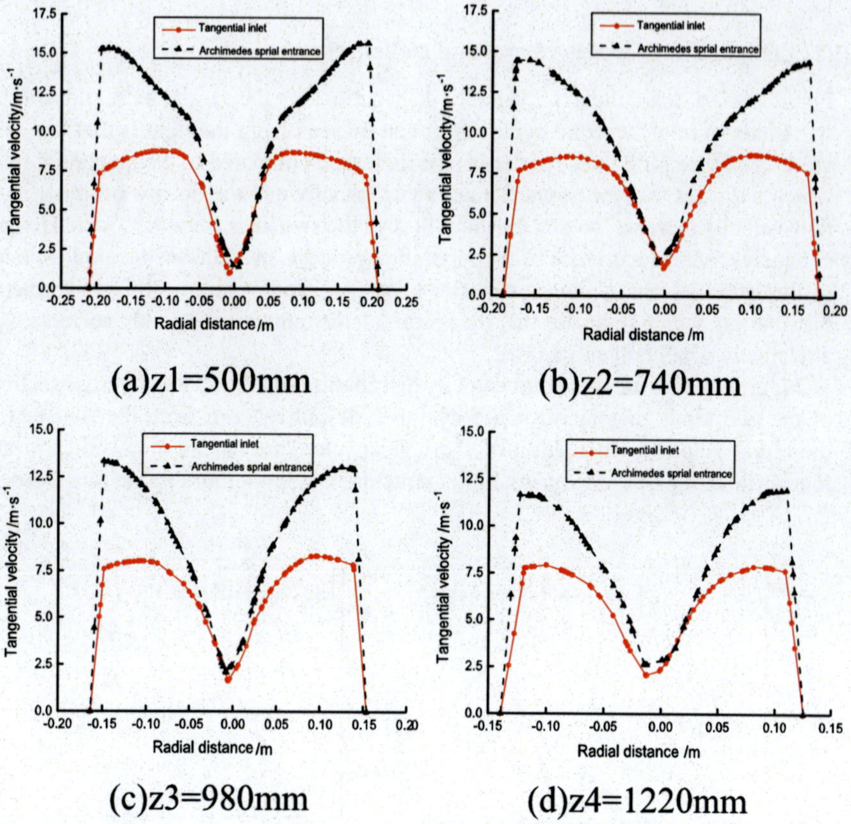

Fig. 6.27 Tangential velocity comparisonof different sections

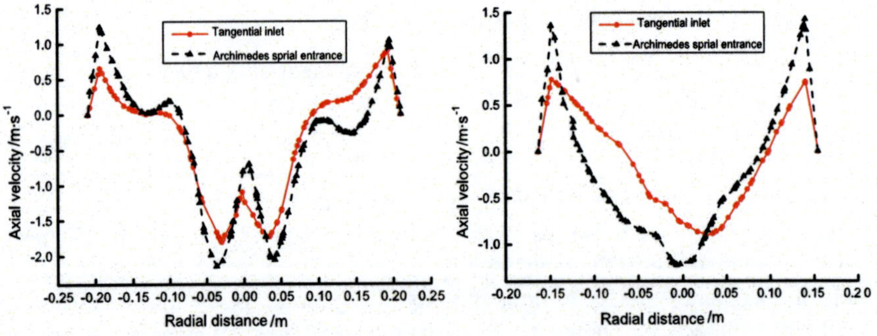

(a) z1=500mm (b) z3=980mm

Fig. 6.28 Axial velocity comparisonof different sections

(2) Comparison of axial velocity and radial velocity

Figures 6.28a, b are the axial velocity distributions on the z1 and z3 sections in the two kinds of inlet structure cyclones. It can be seen from the figures that the axial velocity of the spiral inlet swirler is also increased compared to the tangential inlet, which indicates that the upward inner swirl velocity and the downward outer swirl flow velocity increase, so that the swirl flow in the swirler is smoother, which ismore conducive to the occurrence of the separation process. In addition, the axial velocity of the spiral inlet structure is superior to the tangential inlet in the axisymmetric distribution, which indicates that the spiral inlet is influenced by eddy currents with a relatively small tangential inlet.

Figure 6.29a, b are the radial velocity distributions on the z1 and z3 cross sections of the two kinds of inlet structure cyclones. It can be seen from the figures that the radial velocity distribution of the helical inlet structure is more uniform and regular than that of the tangential inlet structure and has a more stable flow field. In

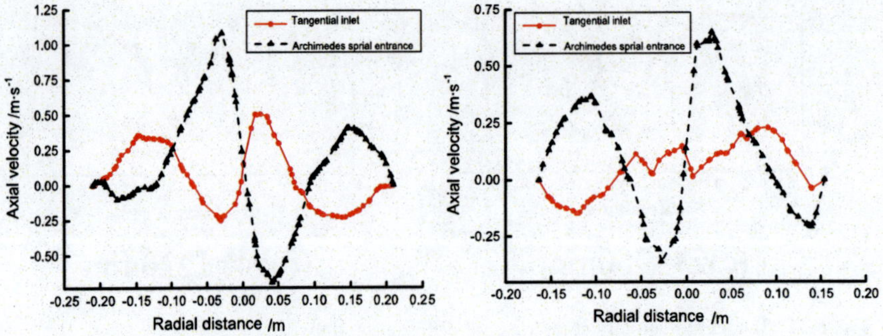

(a) z1=500mm (b) z3=980mm

Fig. 6.29 Radial velocity comparisonof different sections addition, the radial velocity of the helical inlet is also numerically higher than that of the tangential inlet, which indicates that the solid-phase sedimentation resistance is reduced, which is more conducive to the separation of solid particles.

(3) Comparative analysis of volume fraction of solid phase

Figure 6.30 shows the distribution of the volume fraction of the solid phase in the cyclone of the two inlet structures compared with the cloud diagram. Figure 6.31 shows the volume fraction of the solid phase in the sections of the cone z4 and z5. As

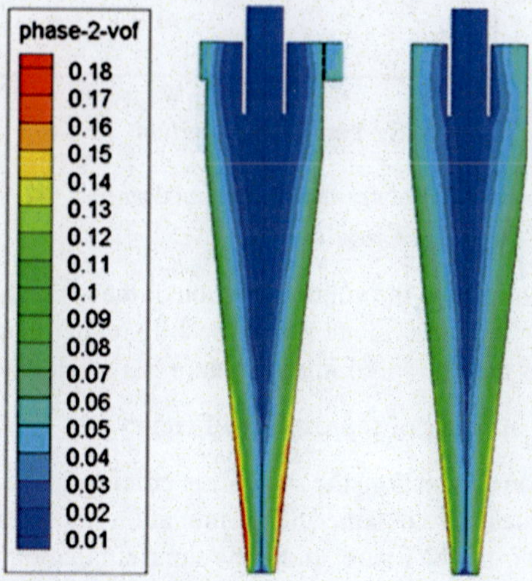

Fig. 6.30 Comparison contours of sand fraction

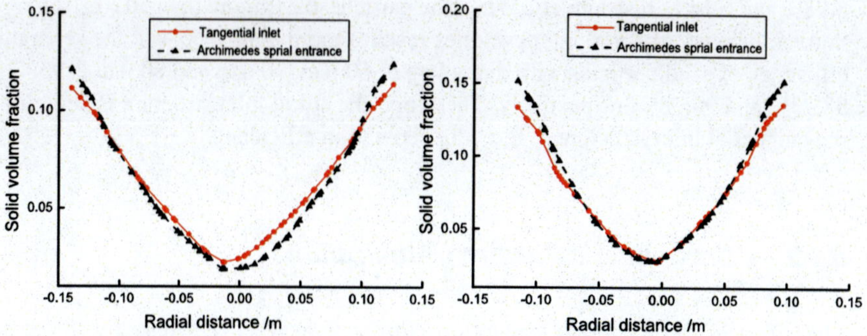

(a) z=1220mm (b) z=1460mm

Fig. 6.31 Sand fraction comparisonof different sections

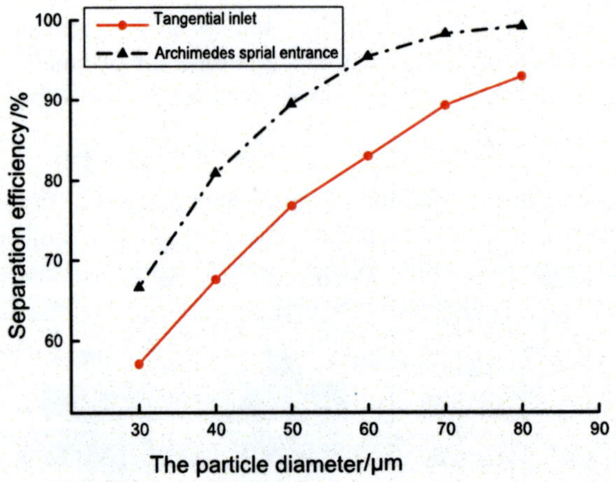

Fig. 6.32 Separation efficiency of the sand in different diameters

can be seen from the figure, the volume fraction of the solid phase in the cone section of the spiral inlet structure cyclone is significantly higher than that of the tangential inlet, which shows that the spiral inlet structure has a better separation effect.

(4) **Comparative analysis of separation efficiency.**

After the simulation calculation, the separation efficiencies of 30, 40, 50, 60, 70, and 80 μm sand particles were obtained. The simulation results are shown in Fig. 6.32.

As can be seen from the figure, under the same simulation conditions, the greater the sand diameter, the higher the separation efficiency of the cyclone. Compared with the tangential inlet structure, the spiral inlet structure increased the separation efficiency of sand particles with different diameters by 9.5%, 13.26%, 12.67%, 11.43%, 8.93% and 6.14%, respectively. When the particle size is more than 50 μm, the separation efficiency of spiral inlet cyclones reaches more than 95%, and the separation efficiency of sand particles with diameters of 60 μm, 70 μm and 80 μm is 95.42%, 98.23% and 99.08%, respectively. Therefore, the spiral inlet structure is superior to the tangential inlet structure in terms of separation efficiency.

6.1.3 Ultrasonic Air Purifier Optimization

The effect of the ultrasonic cavitation purifier is affected not only by the nature parameters of the transducer but also by the hydraulic conditions inside the purifier. Only when the purifier has a certain dead water area or the water flow rate is small, the ultrasonic wave can play a role. Installing baffles in the purifier can block the flow of water and reduce the flow of most of the water. The FLUENT software was used to optimize the flow field in the ultrasonic processor.

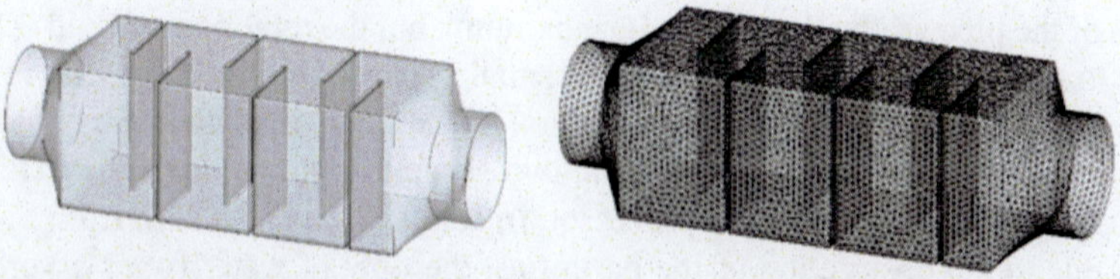

(a) Fluid domain model (b) 3D grid

Fig. 6.33 US cavitation effect to purify the processor model andthe grid fluid domain

1. **Modeling and meshing**

According to the ultrasonic cavitation effect of the designed structure and size of the processor, its geometric model is appropriately simplified, and the increase of water treatment time in the purifier is considered, and the flow direction of the water is adjusted appropriately to increase as much as possible. In the dead water area, the cavitation effect is enhanced, and the fluid domain model is established as shown in Fig. 6.33a; the three-dimensional grid is shown in Fig. 6.33b.

2. **Boundary condition settings**

(1) Inlet boundary: The entrance boundary is defined as the mass flow inlet, The known processing volume is 200 m3/h, and the converted mass flow rate Qm = 55.56 kg/s, turbulence intensity $I = 0.16\left(\mathrm{Re}_{D_H}\right)^{-1/8}$, where D_H is the hydraulic diameter, i.e., the inlet diameter, $D_H = 200$.
(2) Export boundary: defined as free exports (*outflow*);
(3) Wall conditions: Walls with no slip boundaries;
(4) **The water was set as the basic phase**: the density was 998.2 kg/m^3, and the viscosity was 0.001003 Pa·s; the RSM model was used for the steady flow calculation.

3. **Flow field analysis**

Figure 6.34 shows the flow field distribution within the ultrasound processor. From Fig. 6.34a, it can be seen that the fluid enters the cavity through the 6-layer separator and exits through the outlet. During the period along the fluid flow, the cross-sectional

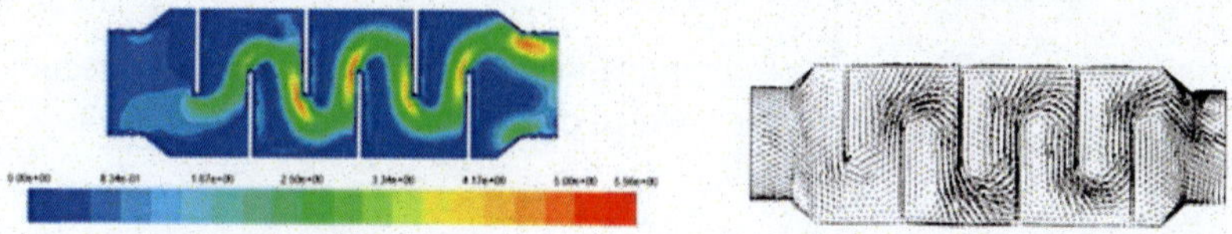

(a) Z=0 velocity profile (b) Z=0 section velocity vector diagram

Fig. 6.34 US cavitation effect purification velocity distribution inside the processor area of the pipe gradually reaches its maximum, but the total flow of fluid is blocked due to the barrier. As the area becomes smaller, part of the fluid velocity increases, and a small area of high-speed flow occurs. At the same time, there are obvious dead zones and low flow areas between the two separators in the processor. This meets the design requirements of the processor. In addition, it can be seen from Fig. 6.34b that due to the obstruction of the partitions, the flow of water forms a vortex flow between every two partitions in the processor and the turbulence intensity increases, which makes the fluid flow through two partitions each time. The spatial position of the fluid is re-arranged between the spaces, thereby increasing the uniformity of the fluid receiving the ultrasonic radiation.

4. **Ultrasonic sound field distribution simulation**

The sound field distribution in the ultrasonic cavitation purifier has a significant effect on its processing effect, and the spatial distribution of the sound field is related to the radiation frequency of the transducer and the structure size of the processor. According to the calculation method of the directivity of the planar array ultrasonic transducer, the radiation sound field distribution of the ultrasonic transducer array in three-dimensional space is calculated.

(1) **Ultrasonic sound field calculation method**

① **Transducer directivity function**

The directivity function of the transducer refers to the spatial distribution function of the radiated sound field or the sensitivity of the receiving array, also called the directional characteristic function, which can describe the spatial distribution of the radiated sound field of a sound source. The directionality function is divided into two types, sound pressure non-normalization and sound pressure normalization, which can be represented by functions $D(\alpha, \theta)$ and $U(\alpha, \theta)$ respectively [12]:

$$D(\alpha, \theta) = |p(\alpha, \theta)|, U(\alpha, \theta) = \frac{|p(\alpha, \theta)|}{|p(\alpha_0, \theta_0)|} \tag{6.22}$$

where:

$|p(\alpha, \theta)|$ The composite sound pressure amplitude in any direction (α, θ)
$|p(\alpha_0, \theta_0)|$ The amplitude of the composite sound pressure in direction (α_0, θ_0) of the maximum of the main beam.

The sound intensity-normalized pointing function can also describe the directional characteristics of the transducer array, which can be represented by $b(\alpha, \theta)$ as:

$$b(\alpha, \theta) = \frac{|I(\alpha, \theta)|}{|I(\alpha_0, \theta_0)|} \tag{6.23}$$

where:

$\lvert I(\alpha, \theta)\rvert$	the sound intensity value in the direction of (α, θ)
$\lvert I(\alpha_0, \theta_0)\rvert$	the sound intensity value in the direction of (α_0, θ_0)

According to the relationship between sound pressure and sound intensity, you can get:

$$I(\alpha, \theta) = \frac{1}{2\rho c}|p(\alpha, \theta)|^2 = \frac{1}{2\rho c}D^2(\alpha, \theta) \tag{6.24}$$

where:

ρ	Medium density;
c	The speed of wave propagation in the medium.

The relation between the sound pressure normalization function and the sound intensity normalization function obtained from Eqs. (6.23)–(6.24) is as follows:

$$b(\alpha, \theta) = D^2(\alpha, \theta) \tag{6.25}$$

② Directivity function of a single circular transducer

For a single point source, set its directivity function to $D_{point}(\alpha, \theta, \omega)$, The sound pressure of the point source at a distance of r can be expressed as:

$$p = K D_{\text{point}}(\alpha, \theta, \omega)\frac{1}{r}e^{j(\omega t - kr)} \tag{6.26}$$

where:

ω	ultrasonic angular frequency;
k	Wavenumber, $k = 2\pi/\lambda$;
K	constant, $K = p_0/\lambda$, p_0 is the initial sound pressure amplitude.

Figure 6.35 is a schematic diagram of the sound field of a circular transducer, with the radius of the vibration end face being a, dividing the end face into n equal areas.s_i Represents the ith area. When $n \to +\infty$, each small vibration surface can be seen as a point source.r_i is the distance between area s_i and far-field P, which is also the sound path of the sound waves emitted from this area to P.r_0 The distance from the area s_0 to P in the center of the vibrating face.Δr_i is a sound path difference of r_0 with respect to r_i, and the vibration amplitude is the same for each area of s_i, so when each small area s_i is used as a point sound source, the constant K in Eq. (6.26) is the same.

For a single circular transducer, the directivity function can be expressed as:

$$D_c(\alpha, \theta, \omega) = \frac{P}{P_{\theta=0}} = \frac{\sum_{i=0}^{n-1} p_0 e^{j\Delta r_i k}}{n \cdot p_0} \tag{6.27}$$

Fig. 6.35 Circular acoustic transducer to coordinate diagram

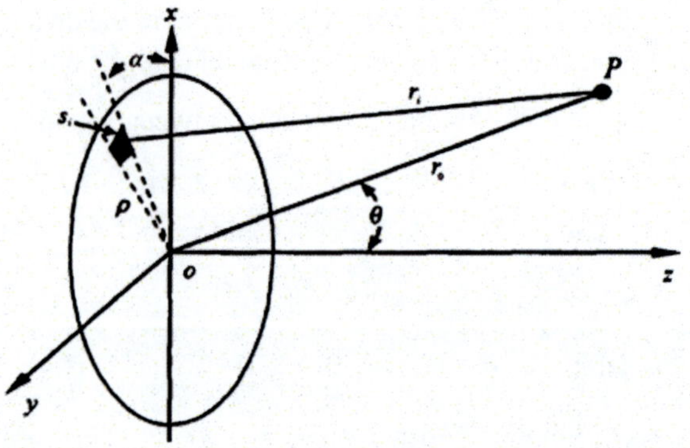

where:

- P the total sound pressure from each approximate point source to the far field;
- $P_{\theta=0}$ the total sound pressure of each approximate point source to the far field when $\theta = 0$

$$P = \sum_{i=0}^{n-1} p_0 e^{j\Delta r_i k} = K\frac{1}{r_0}e^{j(\omega t - kr_0)} \times \sum_{i=0}^{n-1} e^{jk}|\vec{\rho}_i|(\cos\alpha_i \cos\alpha \sin\theta + \sin\alpha_i \sin\alpha \sin\theta) \tag{6.28}$$

As $t \to +\infty$, The summation is converted to integral as:

$$P = K\frac{1}{r_0}e^{j(\omega t - kr_0)} \times \int_0^a \rho\left(\int_\alpha^{2\pi} e^{jk\rho\sin\theta\cos(\alpha_i-\alpha)}d\alpha_i\right)d\rho \tag{6.29}$$

In the above formula, when we integrate α_i for one period, we choose from α to $\alpha + 2\pi$, according to the nature of the Bessel function:

$$P = K'\frac{2\pi a^2}{r_0}e^{j(\omega t - r_0 k)}\frac{J_1(ka\sin\theta)}{ka\sin\theta} \tag{6.30}$$

$$P_{\theta=0} = K'\frac{\pi a^2}{r_0}e^{j(\omega t - kr_0)} \tag{6.31}$$

The directivity function of a circular transducer with a single radius a can be obtained from Eqs. (6.30) and (6.31).

$$D_c(\alpha, \theta, \omega) = \frac{P}{P_{\theta=0}} = \left| 2\frac{J_1(ka \sin\theta)}{ka \sin\theta} \right| \tag{6.32}$$

where:

J_1 first-order Bessel function.

③ Transducer array directivity function.

If the array transducer directivity function is to be derived from the directivity function of a single transducer, the following assumptions are also required:

I $r >> l^2/\lambda$, where r represents the distance from the center of the transducer array to the observation point, $l$is the maximum size of the array, and λ is the wavelength. This condition indicates a far-field condition, in which acoustic waves can be viewed as plane waves;

II The elements of the array transducer have no interference effect on the sound waves in the arrangement;

III The mutual radiation effect between each element of the array transducer is not considered.

Based on the above assumptions, according to formula (6.32), the directivity function for a discrete array three-dimensional space composed of N arbitrarily distributed point sources is obtained:

$$D(\alpha, \theta, \alpha_0, \theta_0, \omega) = \frac{\left| \sum\limits_{i=1}^{N} \tilde{A}_i e^{-j\Delta\varphi_i} \right|}{\left| \sum\limits_{i=1}^{N} \tilde{A}_i \right|} \tag{6.33}$$

where:

$\tilde{A}_i$ the amplitude of the voltage received by the i-th element in the far-field region;

$\Delta\phi_i$ The phase difference between the direction of the i-th array element and the main maximum direction.

Since each array element has the same amplitude, $\tilde{A}_1 = \tilde{A}_2 = \ldots = \tilde{A}_i$, the formula (6.33) can be simplified as:

$$D(\alpha, \theta, \alpha_0, \theta_0, \omega) = \frac{1}{N} \left| \sum_{i=1}^{N} e^{-j\Delta\phi_i} \right| \tag{6.34}$$

For the derivation of the directivity function of complex transducer arrays, the following two theorems can be used for simplification.

I Product theorem: A composite matrix consisting of $N = N_1 N_2$ array elements contains N2 identical subarrays, and each subarray consists of N1 identical vibration elements. The directivity function of the composite array is equal to the directivity of the subarrays. The product of functions $D_1(\alpha, \theta, \omega)$ and $D_2(\alpha, \theta, \omega)$ is as follows.

$$D(\alpha, \theta, \omega) = D_1(\alpha, \theta, \omega) D_2(\alpha, \theta, \omega) \tag{6.35}$$

II Additive Theorem: If the transducer array is composed of different types of subarrays, the directivity function of the composite array can be seen as a linear combination of the directivity functions of each subarray, and therefore for a composite array composed of m different subarrays, its directionality function can be expressed as:

$$D(\alpha, \theta, \omega) = \frac{D_1(\alpha, \theta, \omega)\left|\sum_{i=1}^{n_1} A_{1i}\right| + \ldots + D_m(\alpha, \theta, \omega)\left|\sum_{i=1}^{n_m} A_{mi}\right|}{\left|\sum_{i=1}^{n_1} A_{1i}\right| + \ldots + \left|\sum_{i=1}^{n_m} A_{mi}\right|} \tag{6.36}$$

Based on the above two theorems and the general expression of the directivity function of the transducer array, the directivity function of a line array with N identical array elements can be derived as:

$$D(\alpha, \theta, \omega) = \frac{\sin\left(\frac{\pi N d}{\lambda} \cos\alpha \sin\theta\right)}{N \sin\left(\frac{\pi d}{\lambda} \cos\alpha \sin\theta\right)} \tag{6.37}$$

The directivity function of the N × M complex array composed of M line arrays is:

$$D(\alpha, \theta, \omega) = \frac{\sin\left(\frac{\pi N d_1}{\lambda} \cos\alpha \sin\theta\right)}{N \sin\left(\frac{\pi d_1}{\lambda} \cos\alpha \sin\theta\right)} \frac{\sin\left(\frac{\pi M d_2}{\lambda} \cos\alpha \sin\theta\right)}{M \sin\left(\frac{\pi d_2}{\lambda} \cos\alpha \sin\theta\right)} \tag{6.38}$$

The rectangular array of N × M circular piston transducers is shown in Fig. 6.36.

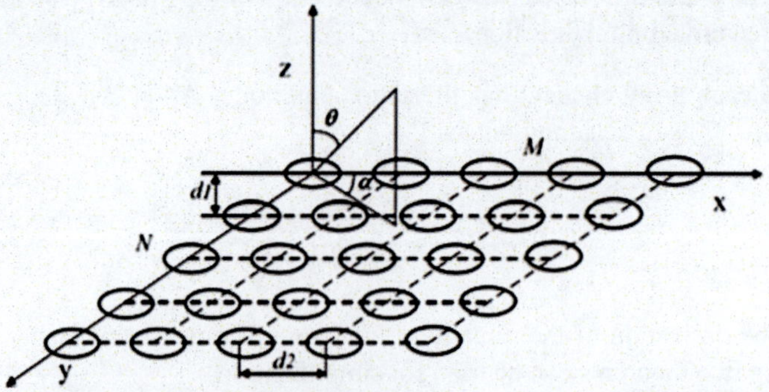

Fig. 6.36 Rounded rectangular array transducer

The directionality functions of the line array consisting of N and M array elements are respectively set. Then, the directivity function of the N × M array can be derived viathe multiplicative theorem and the directivity function of a single circular transducer:

$$\begin{aligned} D(\alpha,\theta,\omega) &= D_N \cdot D_C \cdot D_M \cdot D_C \\ &= \frac{\sin\left(\frac{2\pi a N d_1}{\lambda}\cos\alpha\sin\theta\right)}{N\sin\left(\frac{2\pi a d_1}{\lambda}\cos\alpha\sin\theta\right)} \cdot \frac{2J_1\left(\frac{2\pi a}{\lambda}\sin\theta\right)}{\frac{2\pi a}{\lambda}\sin\theta} \cdot \\ &\quad \frac{\sin\left(\frac{2\pi a M d_2}{\lambda}\cos\alpha\sin\theta\right)}{M\sin\left(\frac{2\pi a d_2}{\lambda}\cos\alpha\sin\theta\right)} \cdot \frac{2J_1\left(\frac{2\pi a}{\lambda}\sin\theta\right)}{\frac{2\pi a}{\lambda}\sin\theta} \end{aligned} \tag{6.39}$$

④ **Transducer array sound pressure calculation.**

According to (6.31), for an N × M rectangular transducer array, the total sound pressure of the transducer at any point (m,n) in the far field region can be expressed as:

$$P_{\theta=0} = M \cdot N \cdot K' \frac{\pi a^2}{r_0} e^{j(\omega t - k r_0)} \tag{6.40}$$

From the Eq. (6.30), the directivity function of the entire array is:

$$D(\alpha,\theta,\omega) = \frac{P}{P_{\theta=0}} \tag{6.41}$$

Then the sound pressure P available at any point can be expressed as a product:

$$P = D(\alpha,\theta,\omega) \cdot P_{\theta=0} = D(\alpha,\theta,\omega) \cdot M \cdot N \cdot K' \frac{\pi a^2}{r_0} e^{j(\omega t - k r_0)} \tag{6.42}$$

The sound pressure amplitude PA at any point is:

$$\begin{aligned} P_A = K' \frac{\pi a^2}{r_0} \frac{\sin\left(\frac{2\pi a N d_1}{\lambda}\cos\alpha\sin\theta\right)}{\sin\left(\frac{2\pi a d_1}{\lambda}\cos\alpha\sin\theta\right)} \cdot \frac{2J_1\left(\frac{2\pi a}{\lambda}\sin\theta\right)}{\frac{2\pi a}{\lambda}\sin\theta} \cdot \\ \frac{\sin\left(\frac{2\pi a M d_2}{\lambda}\cos\alpha\sin\theta\right)}{\sin\left(\frac{2\pi a d_2}{\lambda}\cos\alpha\sin\theta\right)} \cdot \frac{2J_1\left(\frac{2\pi a}{\lambda}\sin\theta\right)}{\frac{2\pi a}{\lambda}\sin\theta} \end{aligned} \tag{6.43}$$

(2) **Matlab based ultrasonic sound field simulation**

The sound field distribution in the ultrasonic cavitation effect purification processor has a significant effect on its functionalities.The sound field in the designed ultrasonic cavitation cleaning processor is simulated by Matlab software. When simulating, the processor can be simplified into a rectangular processor with two upper and lower transducer arrays. First consider only the lower array, using formula (4.94) to find the sound pressure at any point in the processor, then consider the upper array, which can be considered to be the lower array reversed, then at the point of (α, θ, ω), only the angle with respect to the Z axis is different from the upperarray, which can be obtained by the cosine theorem. Using the same method, the sound pressure P_2 of the lower layer array at (α, θ, ω) can be obtained, and the sum of the two is the total sound pressure.

Since the cavitation effect in the liquid is greatly affected by the sound pressure amplitude and the sound energy density uniformity [13], the average value P_{am} of the sound pressure amplitude is used to represent the intensity of liquid sound pressure in the processor, and the non-uniformity index NU_I of the sound field is used to represent the performance of the processor in terms of sound field uniformity and sound energy utilization. The calculation formulas are:

$$P_{am} = \frac{1}{V_0} \iiint_{V_0} |P_A|$$

$$NU_I = \frac{1}{P_{am}} \sqrt{\frac{\iiint_{V_0} (|P_A| - |P_{am}|)^2 dV}{V_0}}$$

The V_0 is the sound field volume. Then, the greater the average value of the sound pressure amplitude, the smaller the sound field non-uniformity index, and the better the performance of the ultrasonic processor in terms of sound field uniformity and sound energy utilization.

According to the above theory, MATLAB software was used to program and study the sound pressure, sound intensity distribution, sound pressure amplitude average value, and non-uniformity index of the liquid in the processor, and compare the energy utilization rate fordifferent structure sizes to determine the processor to determine the final structure size. Please refer to Appendix 1 for the specific procedures.

Figure 6.37 shows the sound field simulation results of the ultrasonic processor, where (a) and (b) are the three-dimensional cross-sectional distribution diagrams of sound pressure and sound intensity in the processor, respectively. (c) and (d) are the graphs of sound pressure and sound intensity distribution, respectively,when the heights are in the *H/2* plane.

According to the simulation results, the maximum value of the sound pressure amplitude in the sound field in the ultrasonic processor is 5.23 × 106 Pa. Substituting the simulation results into the formulas (6.44) and (6.45), the average value of the sound field amplitude is P_{am} 5.05 × 105 Pa. The non-uniformity index *NUI* is 1.2165.

(3) **Influence of cavity height H on the sound field**

The size of the ultrasonic cavity purifier structure has a great influence on the sound field distribution, so the sound field distribution in the liquid of the purifier with different structure sizes is calculated. Taking the height*H* of the purifier cavity (that is, the distance from the transducer) as the simulation object, the variation of the sound field distribution in the processor is studied by fine-tuning *H*. Finally, the

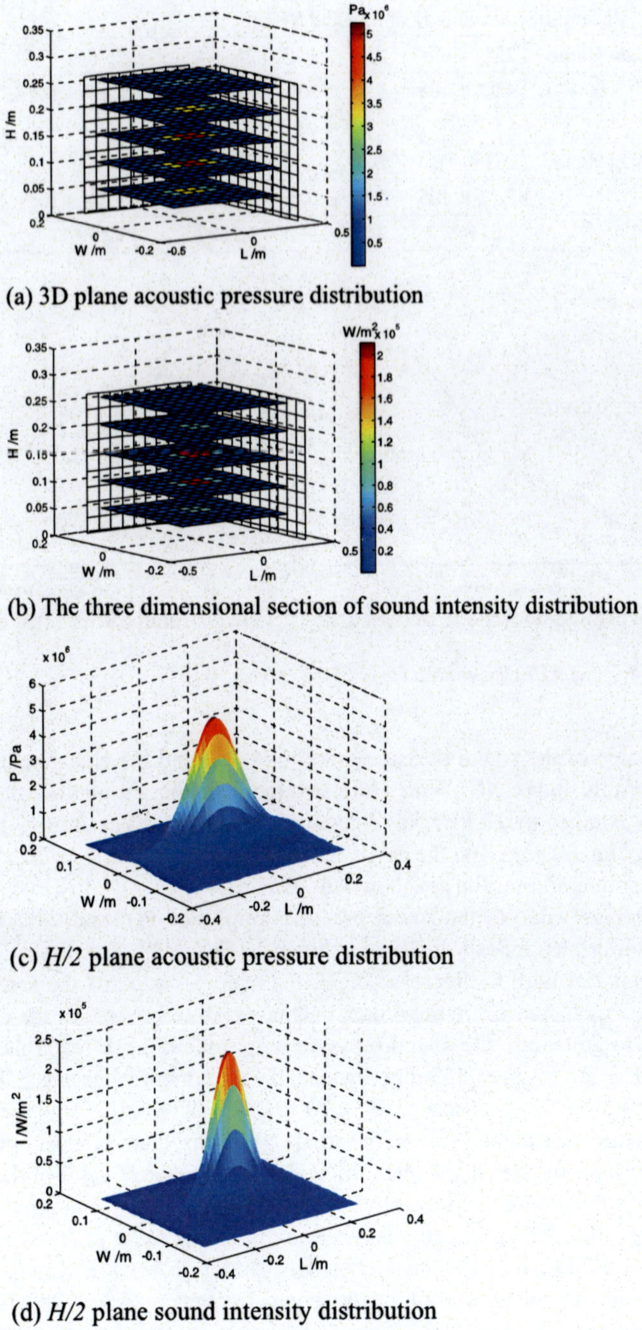

Fig. 6.37 The processor within the acoustic field distribution

Table. 6.2 Under different cavity H of P_{am} and NUI

Cavity height H/mm	251	253	255	257	259
P_{am}/Pa	4.97 × 105	4.98 × 105	5.03 × 105	5.04 × 105	5.08 × 105
NUI	1.2355	1.2324	1.2234	1.2240	1.2165
Cavity height H/mm	261	263	265	267	269
P_{am}/Pa	5.12 × 105	5.23 × 105	5.31 × 105	5.25 × 105	5.19 × 105
NUI	1.2144	1.208	1.2032	1.1992	1.1901

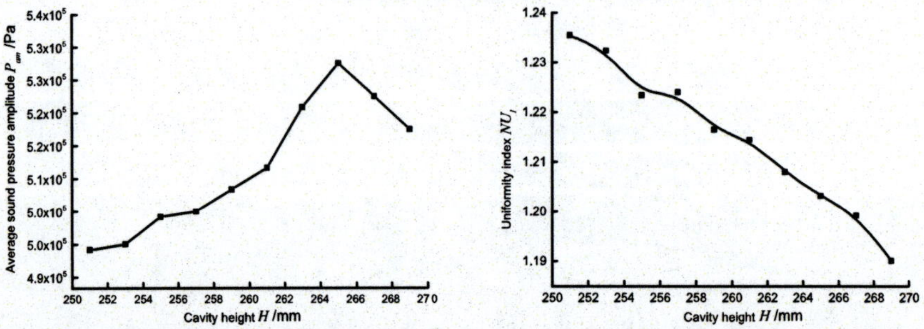

(a)P_{am} change with the H of the curve (b)NUI change with the$H$ of the curve

Fig. 6.38 P_{am} and NUI change with the H of the curve

average value of the sound pressure amplitude P_{am} and the change patterns of the non-uniformity index NUI with $H$are obtained. In the simulation, the ultrasonic frequency is taken as 28 kHz, the length of the cavity is kept at 640 mm and the width is set to 280 mm, and the cavity height changes around $H = 20 \times \lambda/4 = 259$ mm. The simulation results are shown in Table 6.2.

The average value of the sound pressure amplitude P_{am} and the NUI value of the non-uniformity index with H are shown in Fig. 6.38. From the Fig. 6.38a, it can be seen that with the increase of H, the average value of the sound pressure amplitude P_{am} increases first and then decreases. In the course of the change of H from 251 to 269 mm, the sound pressure amplitude did not reach the maximum value at $H = 20 \times \lambda/4 = 259$ mm. Instead, the maximum value was 5.31 × 105 Pa at $H = 265$ mm. This is because $H = 20 \times \lambda/4 = 259$ mm is only the theoretical optimal value in the ideal medium. In actual conditions, there is a loss of the acoustic wave when it propagates in the medium, which leads to the H value of the maximum P_{am} deviating from the theoretical value; From figure (b) we can see that the non-uniformity index NUI gradually decreases with the increase of H, when $H = 265$ mm, $NUI = 1.2032$, which is relatively small. Therefore, for a rectangular ultrasonic cavitation-effect processor with an ultrasonic frequency of f = 28 kHz, a length of 640 mm, and a width of 280 mm, it achieves optimal performance in terms of uniformity and sound energy utilizationwhen $H = 265$ mm.

Table 6.3 P_{am} and *NUI* under different US frequency

Ultrasonic frequency f /KHz	20	30	40	50	60	70	80
P_{am} /Pa	6.18 × 105	5.02 × 105	4.68 × 105	4.42 × 105	4.41 × 105	4.14 × 105	3.7 × 105
NUI	0.8022	1.2901	1.5259	1.6458	1.7041	1.7495	1.767

(4) The effect of ultrasonic frequency on the sound field.

Ultrasonic cavitation effect processor ultrasonic frequency will also have a great impact on its sound field distribution. Therefore, this section will mainly study the average value of the sound pressure amplitude P_{am} and the non-uniformity index *NUI* in the ultrasound cavitation effect processor at different ultrasonic frequencies. During the simulation, the structural dimensions of the processor remain unchanged, and the ultrasonic frequencies f are taken as 20 kHz, 30 kHz, 40 kHz, 50 kHz, 60 kHz and 70 kHz, and 80 kHz, respectively. The simulation results are shown in Table 6.3.

The average value of the sound pressure amplitude P_{am} and the non-uniformity index *NUI* with the ultrasonic frequency is shown in Fig. 6.39. From Fig. 6.39a, it can be seen that the average value of the sound pressure amplitude P_{am} decreases with increasing frequency, indicating that as the frequency increases, the cavitation effect of the liquid in the processor decreases. This is because an increase in the frequency will cause the sound wave expansion time to become shorter and the cavitation nucleus will not grow into a cavitation bubble that can produce an effect. Even if a cavitation bubble is formed, the compression time of the sound wave becomes shorter, resulting in the possibility of cavitation bubbles. It's too late to crash. In addition, the high-frequency ultrasonic energy consumption in the liquid is faster, in order to obtain the same treatment effect, high-frequency ultrasound requires more energy. From figure (b), it can be seen that the non-uniformity index

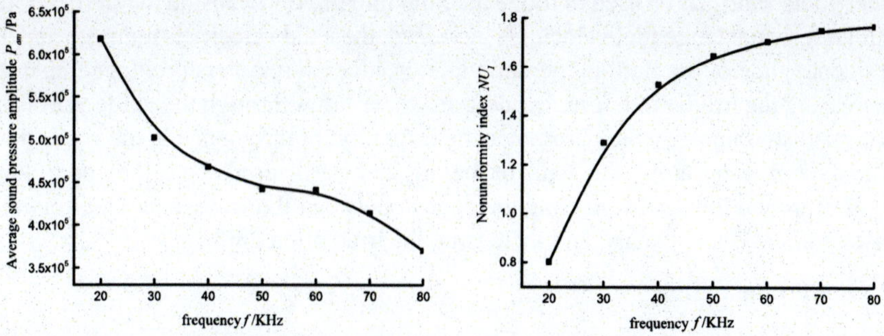

(a)P_{am} along with the change of frequency curve (b) *NU*I along with the change offrequency curve

Fig. 6.39 P_{am} and *NUI* along with the change of frequency curve

NUI increases with the increase of ultrasonic frequency, which shows that as the frequency increases, the processor's acoustic processing uniformity and acoustic energy utilization rate gradually decrease. Therefore, for the rectangular ultrasonic cavitation effect processor discussed in the book, in order to improve the effect of ultrasound, lower frequency ultrasonic waves should be used.

6.2 Optimization of the Sterilization Unit

6.2.1 Optimization of UVC Sterilization Unit

1. **Optimization principle**

The main factors influencing the effect of UV treatment are ballast water quality, microbial species, UVC radiation dose, etc. Among them, UVC irradiation dose can be optimized by changing the UVC sterilizer's own structure. From $D = I \times t$, we can see that in the case of certain light intensity, the longer the residence time of the ballast water in the sterilizer, the greater the irradiation dose, the better the bactericidal effect, and the light intensity is determined by the power of the lamp tube and the lamp tube. The distance from the point is determined, and the residence time is determined by the structure of the sterilizer itself. In addition, some UVC sterilizers have a large difference in the irradiation dose of the fluid in the sterilizer due to the uneven arrangement of the lamps or the structural design is not reasonable. The over-dosed area will cause energy waste, and the area with small dose can not meet the sterilization requirements. Therefore, in the optimization, it should be considered that a relatively uniform radiation dose is obtained in the sterilizer so that the sterilization is more efficient. In summary, the UVC sterilizer is optimized in consideration of the uniformity of the UVC irradiation dose and the fluid residence time: the guide vane is installed at the entrance of the original UVC sterilizer, and the inlet guide vane can be used to rotate the fluid into the UVC. Inside the sterilizer, the circumferential tangential velocity is increased, the axial velocity is reduced, and the residence time of the fluid in the ultraviolet irradiation region is prolonged, thereby increasing the irradiation dose. In addition, water flows through the guide vanes and creates disturbances, which promotes the formation of turbulence in the water flow. The flow of water along the walls of the original sterilizer also has the opportunity to flow around the lamps, making the contact between the water and the ultraviolet rays more fully and evenly, and increasing the irradiation uniformity.

2. **Guide vane design**

Referring to the three-dimensional design theory of the axial-flow pump impeller blade, the guide vane is regarded as an impeller with a rotation speed of 0 r/min, and the blade section diameter d_i, the number of blades z, the blade thickness δ, and the installation angle β are designed as key parameters. The initial design parameters of

Table. 6.4 Initial design parameters of diversion vane

Blade section diameter/mm	Maximum root thickness/mm	The maximum thickness of the leaf/mm	Number of leaves/sheet	Mounting angle/°
200	12	7	3	35

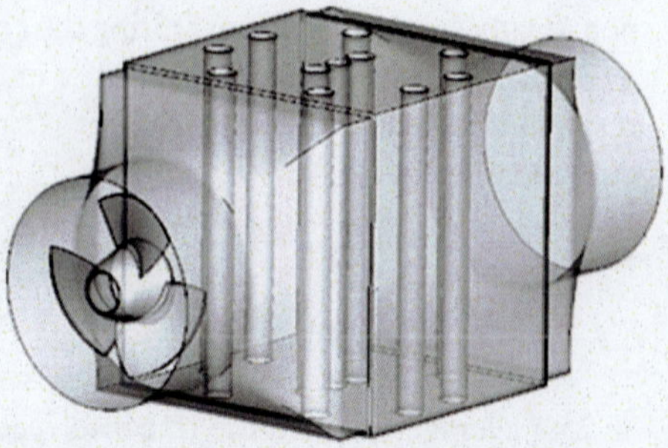

Fig. 6.40 UVC sterilizer with guide vane

the guide vanes are shown in Table 6.4. The structure of the UVC sterilizer equipped with guide vanes is shown in Fig. 6.40.

3. **Simulation of flow field and residence time in UVC sterilizer**

(1) **Simulation method**

At present, scholars often use impulse trace transients to solve the additional component transport equations to simulate the residence time distribution in a certain system. Therefore, the flow field and fluid residence time in the sterilizer were numerically simulated using the RSM turbulence model in FLUENT software and the Species Transport model. In order to facilitate the observation of the residence time and flow field distribution in the sterilizer, the four cross sections of $x = 0$, $z = 0$, interface 1, and interface 2 were set as the monitoring surface, as shown in Fig. 6.41.

The simulation of the residence time is as follows.The inlet of the sterilizer adopts a mass flow inlet, the mass flow rate is $Q_m = 55.56$ kg/s, the outlet is a free outlet,

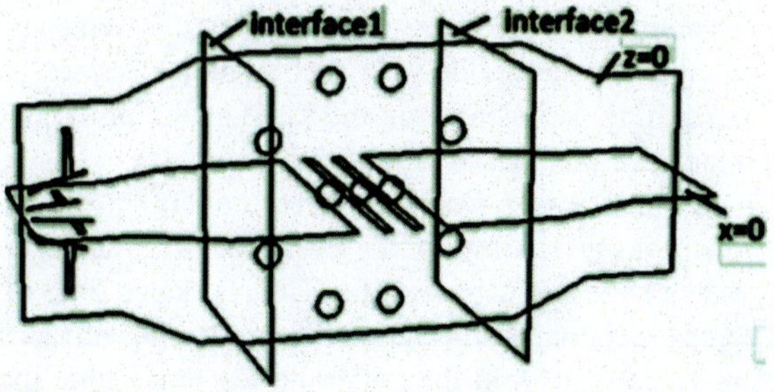

Fig. 6.41 Monitoring surface setup diagram

and the wall surface is selected for no-slip conditions. Considering gravity, a simple solution is adopted. The simulator performs a simulation calculation. First, the test fluid water was introduced into the reactor, and the RSM model was used to simulate the basic flow field under steady state conditions. The fluid flow field was calculated. After the convergence is calculated, the Species Transport model is activated, 1% tracer is injected into the imported components of the boundary conditions (its physical parameters are the same as that of water), and the model is changed for non-steady state calculation. During the calculation process, the tracer concentration c on interface1, interface2, and outlet are monitored separately with respect to the residence time t. Finally, the average residence time $\bar{t}$ of the tracer on each plane is calculated using Eq. (6.46):

$$\bar{t} = \frac{\sum_i t_i c_i}{c_i} \tag{6.46}$$

The average time for the tracer to reach the entrance surface (interface1) of the ultraviolet lamp is denoted by t_1, and the average time for reaching the exit surface (interface2) of the cavity is denoted by t_2, the average of the fluid in the cavity of the ultraviolet lamp. The residence time $\bar{t}$ is $t_2 - t_1$. The average time that the tracer reaches the outlet surface of the sterilizer is the total residence time of the fluid in the sterilizer. By simulating the original UVC sterilizer, the average residence time of the fluid in the lumen of the UV lamp was found to be 0.178 s, and the total residence time in the sterilizer was 1.384 s.

(2) **Influence of guide vane structure parameters on residence time**

A variety of guide vane structures were designed according to the section diameter, number of vanes and installation angle of the guide vanes. Based on the simulation method of residence time, the influence of different structural parameters on the residence time of the fluid in the sterilizer was studied.

1. **Guide blade section diameter.**

The cross-section diameter of the guide vane is too large to affect the fluid flow at the inlet. If the cross-section diameter is too small, the impact on the flow in the sterilizer will be small. As a result, it cannot play the role of the guide vane and cannot achieve the expected results. Therefore, for a structurally determined sterilizer, there will be an optimum guide vane section diameter. Based on the sterilizer inlet diameter of 260 mm, the design guide vane section diameters are 175, 200, 225 and 250 mm. The remaining parameters are the same as the initial design parameters. The simulated residence time curve is shown in Fig. 6.42.

As can be seen from the figure, the RTD curves show a unimodal distribution, but the peaks and valleys of the residence time distribution curves of the different guide vanes have slightly changed. The residence time of the guide vanes with different cross-sectional diameters on the monitoring surfaces of the UVC sterilizer is shown in Table 6.5. From the table, the residence time increases first and then decreases

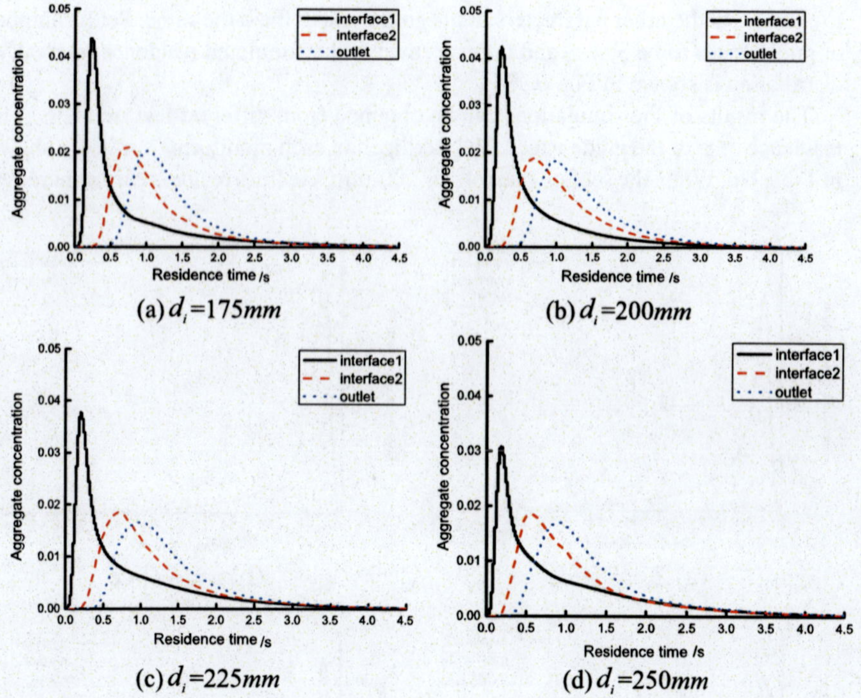

Fig. 6.42 The residence time distribution of the fluid flow in the reactor with different cross sections

Table. 6.5 The residence time distribution of the fluid flow in the reactor with different cross sections.

Section diameter/mm	Interface1/s	Interface2/s	Outlet/s	Average residence time/s
175	0.92	1.222	1.478	0.302
200	0.766	1.225	1.485	0.459
225	0.907	1.243	1.479	0.336
250	0.965	1.24	1.472	0.275

with the increase of the cross-section diameter, and when the section diameter is 200 mm, the average residence time of the fluid in the UV lamp cavity is the longest, which is 0.459 s.

4. **Influence of number of guide vanes on stoppage time**

The excessive number of guide vanes not only fails to fill the UVC sterilizer chamber evenly with the incoming fluid, but also disturbs the fluid flow in the device. Excessive number of guide vanes will affect the effective cross-sectional area of the inlet, which will affect the treatment flow. Therefore, for a particular sterilizer configuration there will be a suitable number of guide vanes. Take the section diameter as $d_i = 200$ mm to ensure that the other parameters of the guide vane remain the same. Setthe number of guide vanes to be 3, 4, 5 and 6, respectively. The simulated residence time curve distribution is shown in Fig. 6.43.

The results of the simulations can be obtained from different leaf numbers. The residence time of the guide vane UVC sterilizer on each monitoring surface is shown in Table 6.6. When the section diameter is 200 mm, the fluid residence time increases

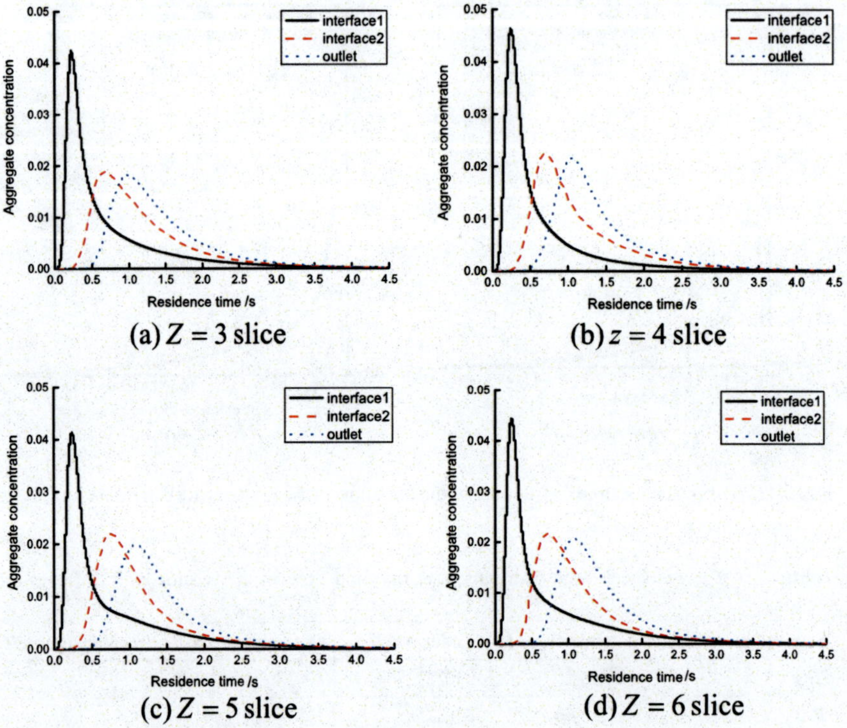

Fig. 6.43 The residence time distribution of the fluid in the reactor with different vanes

Table. 6.6 The retention time of the fluid on the monitoring surfaces of the vanes with different number of vanes in the sterilizer

Number of leaves/sheet	Interface1/s	Interface2/s	Outlet/s	Average residence time/s
3	0.766	1.225	1.485	0.459
4	0.765	1.25	1.486	0.485
5	0.799	1.196	1.485	0.397
6	0.817	1.185	1.477	0.368

first and then decreases with the increase of the number of guide vanes. When the number of guide vanes is 4, the average residence time of the fluid in the UV lamp chamber is the longest, 0.485 s.

5. **Effect of blade mounting angle on stoppage time**

The guide vane installation angle refers to the angle between the chord of the blade root section and the radial section. Excessive installation angle of the blade will make the radial velocity of the fluid entering the sterilizer too large and the axial velocity too small, making the fluid flow too small; If the installation angle is too small, the axial velocity will be too large, and the residence time of the fluid will be too small to function as a guide vane. Therefore, there is a reasonable installation angle for a specific sterilizer structure. When the diameter of the section is $d_i = 200$ mm and the number of guide vanes is 4;the installation angles are 20, 25, 30, 35, and 40 degrees, respectively; and other design parameters are the same,the simulated residence time curve is shown in Fig. 6.44.

For the simulation results, the residence time on the surface of the guide vane UVC sterilizer at different installation angles is shown in Table 6.7.

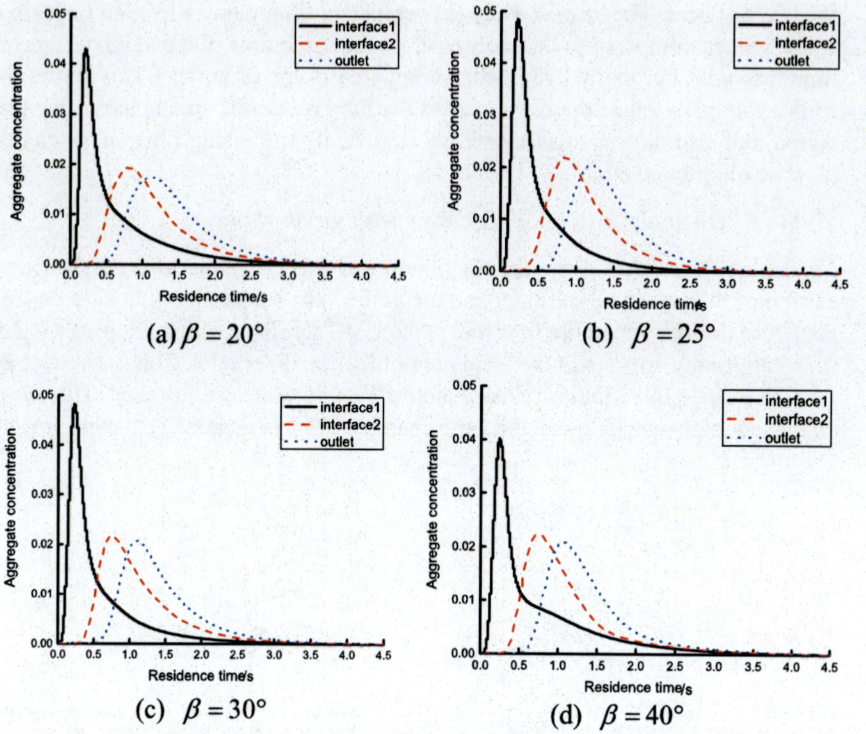

Fig. 6.44 The residence time distribution of the fluid flow in the reactor with different installation angles

Table. 6.7 In the sterilizer, the flow of the fluid on the monitoring surface is controlled by the different guide angles

Mounting angle/°	Interface1/s	Interface2/s	Outlet/s	Average residence time/s
20	0.682	1.189	1.486	0.507
25	0.545	1.158	1.482	0.613
30	0.609	1.178	1.483	0.569
35	0.765	1.25	1.486	0.485
40	0.819	1.173	1.479	0.354

It can be seen from the table that when the section diameter is 200 mm and the number of blades is 4, the residence time of the fluid increases first and then decreases with the increase of the installation angle of the guide vane. When the installation angle is 25°, the fluid is in the UV lamp chamber. The average residence time in the body was the longest, 0.613 s.

In summary, the simulation results show that in the designed guide vane model, when the section diameter is 200 mm, the number of blades is 4 and the installation angle is 25°, the fluid has the longest residence time in the UV-illuminated area of the UV sterilizer. The longest average overcurrent time was 0.613 s, an increase of 0.435 s when compared to the prototype. At the same time, the overall overcurrent time increased by only 0.098 s when compared to the prototype. This shows that adding the guide vane not only increases the fluid residence time in the strong light region, but also have a smallimpact on the overall processing time, achieving the purpose of optimization.

(5) **Flow field analysis in a UV sterilizer with guide vanes.**

Figures 6.45 and 6.46 show the distribution of the flow field at the z = 0 cross section of the prototype sterilizer and the guide vane sterilizer. It can be seen from the figure that the maximum flow rate in the UVC sterilizer with guide vanes is 2.59 m/s, but it only appears in the local area of the sterilizer inlet. This is because the flow area of the inlet fluid is reduced after the guide vanes are installed. There is no significant high-speed flow in the lamp chamber area. In general, as the guide vanes

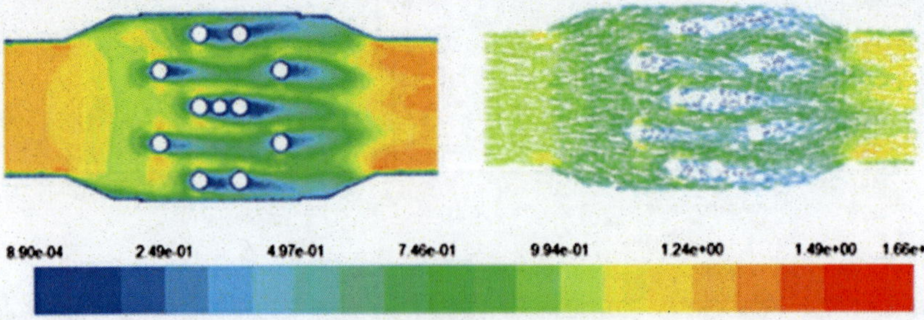

Fig. 6.45 Velocity field distribution in a prototype device

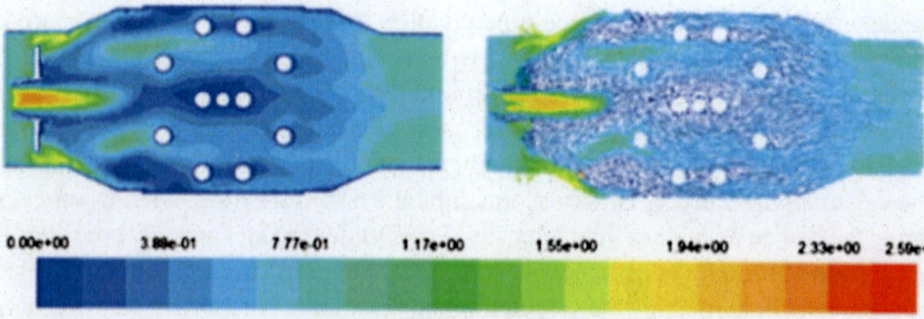

Fig. 6.46 Velocity and flow field distribution in the sterilizer with guide vane

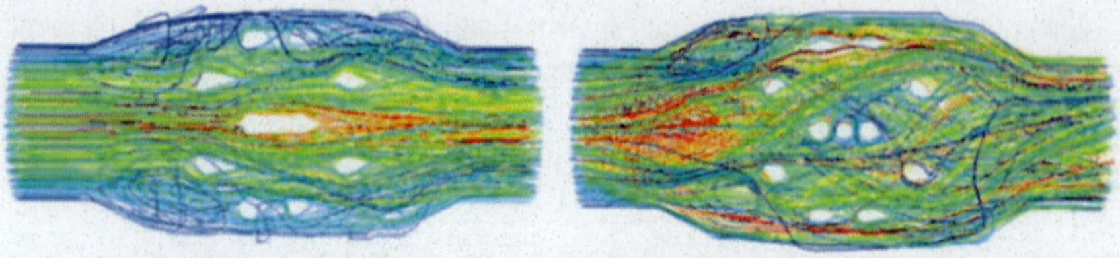

(a)Prototype sterilizer (b) Sterilizer with guide vane

Fig. 6.47 Flow chart of sterilizer

have a blocking effect on the flow of water, the majority of the water flow in the lamp chamber decreases in speed, which increases the UV irradiation time, especially the difference in the velocity of the fluid around the lamp, resulting in a flow field. The distribution of speed is more uniform.

Figure 6.47 is a streamline diagram inside the sterilizer. As can be seen from the figure, the flow lines in the prototype sterilizer are almost linearly distributed except around the lamp. At the same time, due to sharp changes in the cross section of the sterilizer, the speed loss is large, and the angle of the sterilizer appears obvious vortex; After the fluid enters the optimized UVC sterilizer, the diversion of the guide vanes causes a significant disturbance within the sterilizer. The streamline enters the interior of the sterilizer in a rotating state. This causes the flow of water along the wall of the original sterilizer to also flow. Opportunities flow around the lamp, making the contact between the water stream and the UV more adequate and uniform, increasing the uniformity of the illumination. In addition, the swirling phenomenon at the corner of the sterilizer was improved after the addition of the guide vanes.

(3) **Simulation based on FLUENT UV light intensity and dose**

1. **Simulation of UV light intensity**

Ultraviolet light sources are mostly cylindrical, and mathematically, column-type light sources are often assumed to be a collection of multiple point sources, and light radiates from these points in all directions. The light intensity calculation model includes multi-point source superposition (MPSS), multi-segment superposition (MSSS), line source integration (LSI), and column source (Cylinder). The MPSS model divides a linear lamp tube into n point light sources for superposition calculation; The LSI model ignores the absorption of ultraviolet light by the medium. It abandons the MPSS model's limitation on the number of point sources and improves the calculation accuracy. However, this model has certain limitations when calculating the device with weak light intensity. The Cylinder model not only considers the absorption of ultraviolet rays, but also considers the reflection and refraction effects of ultraviolet light, making the calculation more accurate [14]. Based on the analysis above, this chapter selected Cylinder model and MPSS model to simulate the light intensity.

The Cylinder model divides the UVC tube into a number of columnar ring elements, while the MPSS model approximates each cylindrical ring element as a multi-point light source to perform superposition calculation, taking into account the light refraction, reflection, and absorption of the medium to light. The formula for calculating the light intensity distribution is as follows.

$$I(x, y, z) = \frac{P}{4\pi^2 L} \int_{x_0}^{x_1} \int_{\partial_0}^{\pi-\partial_0} \frac{\exp\left(-\lambda\sqrt{(x - x_p)^2 - 2r\sin\theta\sqrt{y_p^2 + z_p^2} + r^2 + y_p^2 + z_p^2}\right)}{(x - x_p)^2 - 2r\sin\theta\sqrt{y_p^2 + z_p^2} + r^2 + y_p^2 + z_p^2} d\theta dx \quad (6.47)$$

where:

P	lamp effective output power;
L	lamp arc length;
θ	reflection angle;
λ	average light transmittance of the solution;
r	lamp tube radius;
x_0, x_1	axial start and end coordinates of the lamp, $x_1 - x_0 = L$;

Based on the mathematical calculation formula of light intensity, the discrete coordinate radiation model (DO) in FLUENT software was used to simulate the light intensity distribution in the ultraviolet sterilizer. The fluid boundary conditions are the same as the boundary conditions set in the flow field simulation. The specific settings of the radiation boundary conditions are as follows: The cylindrical surface of each lamp is divided into 8 parts equally, and 8 radiation directions are set; set the lamp wall type as the semi-transparent translucent wall, wall material is set to quartz, the light intensity on the quartz sleeve surface is calculated according to Lambert's lawto be 10992.8 W/m^2; Since the light intensity simulated by the sterilizer is limited to the UVC section, the light with a wave length of 254 nm is set as the main wavelength band. The settings of other simulation parameters are shown in Table 6.8.

According to the actual sterilizer modeling, the distribution of light intensity in the UVC sterilizer was calculated based on the above simulation method as shown in Fig. 6.48.

Table. 6.8 Analog parameters settings

Fluid water	Density /kg·cm^{-3}	998.2
	Absorption coefficient /m^{-1}	23.26
	Refractive index	1.372
Quartz tube	Density/kg·cm^{-3}	2200
	Absorption coefficient /m^{-1}	44.62
	Refractive index	1.48
	Transmittance	0.9

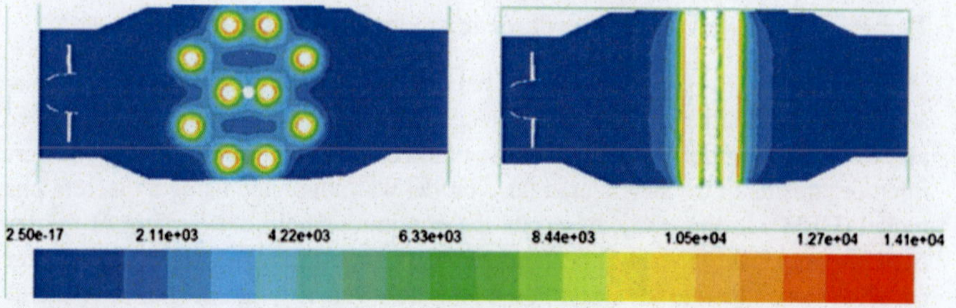

Fig. 6.48 Light intensity distribution of z = 0 section

It can be seen from the figure that the UV intensity distribution inside the lamp cavity is relatively uniform, and the distribution rule of the light intensity is that the ultraviolet light around the lamp area is strong, the maximum intensity is 14065.8 W/m^2, and the area between each lamp is ultraviolet. The intensity is high, the average intensity is above 6330 W/m^2. The weaker areas of the sterilizer appear near the inlet of the reactor wall and the sterilizer lumen, and the farther away from the tube, the weaker the intensity. In addition, the ultraviolet radiation intensity at both ends of the lamp chamber is small, and therefore, the dose of ultraviolet radiation to the ballast water flowing through this region is relatively small. By adding guide vanes at the inlet of the UV treatment device, the velocity of most of the water flow is forcibly reduced, and the flow of water is directed to a region where the UV irradiation intensity is high, and the residence time of the water flow in a region with a higher UV irradiation intensity is increased, which is advantageous to increase the water flow uniformity and concentration.

2. **Numerical simulation of ultraviolet dose**

In the continuous flow sterilizer, as the microorganisms are smaller and the density is similar to that of water, it can be assumed that the flowability of the microorganisms is good and the flow characteristics in the sterilizer are the same as those of the fluid. The trajectory equation of microorganisms in the sterilizer can be Expressed as:

$$\frac{dx_i}{dt} = u_i \tag{6.48}$$

Then, the radiation dose equation that microorganisms receive during fluid flow is:

$$D = \int_0^T I(u_i)dt \quad (6.49)$$

where:

$I(u_i)$ The intensity of radiation at a point at t;
T The total residence time of the microorganism in the sterilizer.

From the Eq. (6.49), it can be seen that the flow path of different microorganisms and the residence time in the sterilizer are all different, so the radiation dose received is also different. In order to obtain the dose of ultraviolet radiation received by different microorganisms in the sterilizer, a dose model was simulated using a discrete phase model (DPM). In the study, based on the continuous field of coupling of fluid and light intensity, the DPM model was activated, the inlet was set as a surface jet source, particles were added to the water at the inlet flow velocity, and the points in the flow direction in the sterilizer were tracked, and the total residence time of the particles was detected at the outlet. Finally, in the post-processing mode, the average radiation intensity I received by each particle in each time step Δt is recovered, and the radiation dose ($I \times \Delta t$) received by each particle in this time step is calculated, and the total can be derived accordingly. The dose within a step can give the total radiation dose received by the particles. The trajectory of the particle inside the sterilizer was simulated as shown in Fig. 6.49.

The statistical analysis method was used to obtain the percentage of radiation dose in different intervals, as shown in Fig. 6.50.

It can be seen from the figure that the dose in the prototype UVC sterilizer is mainly distributed within the range of 100–300 mJ/cm^2, and the average UV Dose is 175.2 mJ/cm^2; The dosage in the guided leaf-type UVC device was mainly distributed in the range of 200 to 400 mJ/cm^2, and the average UV Dose was 259.5 mJ/cm^2. The simulation results showed that the particles were increased by 48.1% when compared to the average value of the UV irradiation dose after the UVC treatment device was modified before and after.

(4) Simulation of ultraviolet light intensity and dose based on MATLAB

1. UV light intensity and dose calculation principle

The basic models used for UVC intensity calculation are Radial model and Point Source Summation model. According to the design characteristics of the reactor shape, Radial radiation model was selected to calculate the radiation intensity and dose inside the sterilizer. The Radial model treats the UV lamp as a linear light source with evenly distributed energy, and the following assumptions are made:

I. UV lamp radiation energy is evenly distributed along the axial direction;

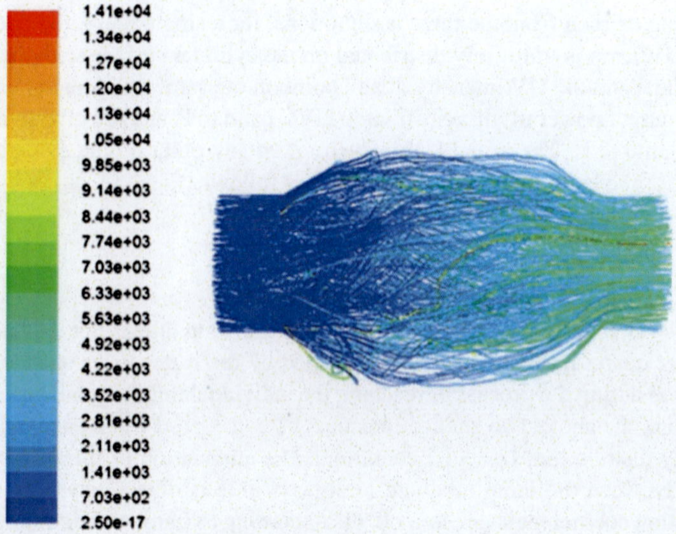

Fig. 6.49 Trajectory diagram of sterilizer particles

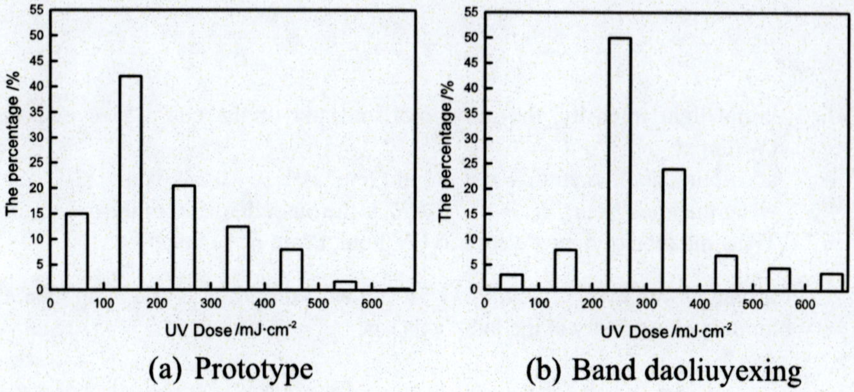

Fig. 6.50 Ultraviolet dose distribution

II. The ultraviolet rays are perpendicular to the axis of the lamp tube and the outer wall of the quartz tube and propagate energy radially outward in the form of a cylinder;

III. Only the absorption of ultraviolet light by contaminants in water is considered, regardless of the refraction of the air, quartz and water interfaces and the scattering and reflection of contaminants in the water.

First calculate the average residence time of the ballast water in the cavity, that is, the UV irradiation time, take the Fluent simulation value of 0.613 s.

As can be seen from the three assumptions, for a single lamp, the output energy of the UV lamp is uniformly distributed on the cylinder surface with the lamp tube as the axis, and the UV intensity at any point on the same cylinder surface is equal. Set the output power of ultraviolet lamp UVC band to P, and the effective arc length of UV lamp to L. The initial UV intensity I_0 at any point on the cylinder from the radius r of the lamp tube can be calculated as follows.

$$I_0 = \frac{P}{2\pi r L} \tag{6.50}$$

However, the ultraviolet energy emitted by the lamp tube is not only absorbed by the outer quartz glass tube but also absorbed by the water layer outside the quartz glass tube during the process of reaching the individual microorganisms in the water. Therefore, the above two sttenuationsmust be considered when calculating the UV intensity distribution. In actual conditions, the attenuation of ultraviolet energy in the air between the lamp tube and the quartz tube is objectively present, and the attenuation coefficient is generally 0.99. According to Lambert's law, the ultraviolet light intensity I at a distance d_w from the lamp water layer can be expressed as:

$$I = I_0 \eta_q e^{-\alpha_w d_w} \tag{6.51}$$

where:

- I_0 initial light intensity, the ultraviolet intensity of the quartz sleeve surface (W/cm^2);
- α_w absorption coefficient of water to light (cm^{-1}), $\alpha_w = 0.2326\ cm^{-1}$;
- d_w water thickness (cm), $d_w = r - R_0$, R_0 is the outer diameter of the quartz tube;
- η_q Transmittance of quartz sleeve to UV light, taken $\eta_q = 0.9$.

From the formulas (6.50) and (6.51), it can be deduced that for a lamp tube, the UV illumination intensity at the lamp axis r is:

$$I = \frac{P\eta_q}{2\pi r L} e^{-\alpha_w (r - R_0)} = \frac{P\eta_q e^{\alpha_w R_0}}{2\pi L L} e^{-\alpha_w r} \tag{6.52}$$

The UV irradiation dose at this point is the following:

$$Dose = I \cdot t = t \cdot \frac{P\eta_q e^{\alpha_w R_0}}{2\pi r L} e^{-\alpha_w r} \tag{6.53}$$

In the Radial model, the UVC intensity at each point in the sterilizer depends only on the radial distance r at this point and is independent of the axial coordinate. When there are several UVC lamps in the reactor evenly distributed along the ring, the total light intensity at any point in the reactor is the accumulated light intensity of all UVC lamps at this point. Therefore, for n lamps, the total irradiation dose at

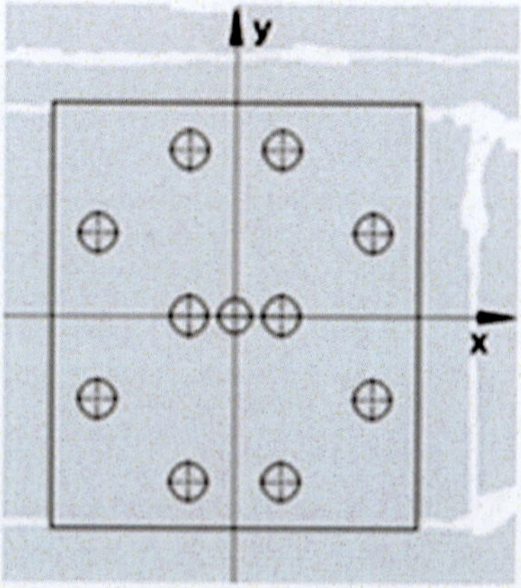

Fig. 6.51 Cross section of sterilization lamp tube chamber

this point is the sum of the irradiation doses at each point from each UV lamp, that is:

$$Dose_t = t \cdot \sum_{i=1}^{n} I_i = t \cdot \sum_{i=1}^{n} \frac{P\eta_q e^{\alpha_w R_0}}{2\pi r_i L} e^{-\alpha_w r_i} \tag{6.54}$$

Since the UVC intensity in Radial model is independent of the axial coordinate, the light intensity distribution rule of each section is exactly the same. Therefore, the average light intensity of the sterilizer is equal to the average light intensity of any section. So, a cross-section inside a sterilizer can be studied, as shown in Fig. 6.51.

According to the above formula of light intensity and dosage, MATLAB software combined with VC ++programming was used to simulate the light intensity and dose in the sterilizer chamber. To calculate the light intensity of each point on any plane in the sterilizer, we must first know the coordinates of each point, and then solve the light intensity of each point according to the formula. Therefore, this study first started from writing VC++ programs, taking points from the crosssection, outputting coordinates (x, y), and calculating the distance r from each point to each lamp. After that, calculation programs in MATLAB were written according to the calculation formula of light intensity and dose. The coordinate output program along withthe light intensity and dose calculation programs are shown in Appendix 2.

In the calculation, the operating results of the VC++ program are imported into MATLAB, and the light intensity and dose calculation procedures are used to obtain the light intensity and dose at each point on the cross section, and the plane light intensity and dose distribution diagram are plotted. The entire flow for calculation is shown in Fig. 6.52.

The output of the coordinate output program is shown in Fig. 6.53:

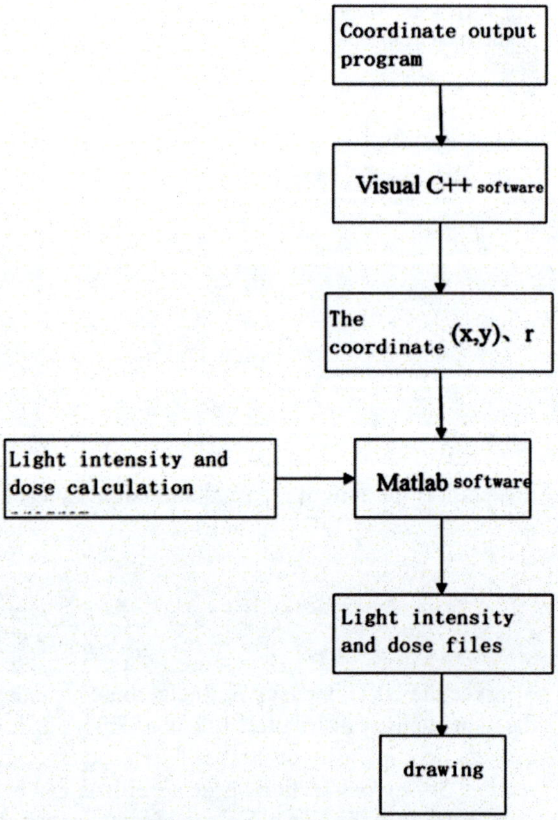

Fig. 6.52 Light intensity and dose calculation

2. Simulation results analysis

The calculated simulation results are shown in Figs. 6.54 and 6.55, respectively, plottedas a plan view and three-dimensional distribution of dose and illumination intensity on any section in the UVC cavity.

From the radiation dose distribution, it can be seen that the radial UV dose distribution in the reaction chamber is relatively uniform. Due to the superposition of radiation from each UV lamp, the dose of UV radiation in the area around the lamp tube is relatively high, and the dose is gradually reduced as the distance from the lamp tube becomes longer; the quartz tube wall is closer to the UV lamp and UV decay is reduced, so the UV radiation dose is the highest. The maximum UV irradiation dose is about 944.1 mJ/cm^2. The dose of UV radiation is relatively low around the cavity of the UV device, and the radiation dose near the entrance of the cavity and the baffle is relatively small, and the minimum dose is about 21 mJ/cm^2. The specific values of the simulation results are shown in Table 6.9.

Figure 6.56 shows the distribution of UVC dose contours.

	A	B	C	D	E	F	G	H	I	J
	r1	r2	r3	r4	r5	r6	r7	r8	r9	r10
	Number	Number	Number	Number	Number	Number	Number	Number	Number	Number
1	37.7359	37.5366	26.9072	20.3961	30.0832	12.6491	21.6333	34.1760	11.7047	25.3180
2	37.4994	37.2022	26.5745	19.9550	29.6648	12.2230	21.3870	34.0206	11.5587	25.2508
3	37.2667	36.8702	26.2452	19.5144	29.2474	11.7987	21.1473	33.8705	11.4285	25.1915
4	37.0381	36.5407	25.9195	19.0741	28.8309	11.3764	20.9146	33.7257	11.3147	25.1401
5	36.8136	36.2138	25.5977	18.6344	28.4155	10.9563	20.6891	33.5863	11.2178	25.0966
6	36.5932	35.8896	25.2797	18.1951	28.0011	10.5386	20.4710	33.4524	11.1383	25.0612
7	36.3771	35.5681	24.9658	17.7564	27.5879	10.1237	20.2606	33.3240	11.0766	25.0338
8	36.1652	35.2494	24.6561	17.3183	27.1758	9.7120	20.0580	33.2012	11.0328	25.0144
9	35.9578	34.9337	24.3508	16.8808	26.7649	9.3038	19.8635	33.0841	11.0073	25.0032
10	35.7548	34.6208	24.0500	16.4439	26.3553	8.8996	19.6775	32.9728	11.0001	25.0001
11	35.5563	34.3111	23.7539	16.0078	25.9471	8.5000	19.5000	32.8672	11.0114	25.0050
12	35.3624	34.0044	23.4628	15.5725	25.5402	8.1057	19.3314	32.7674	11.0409	25.0180
13	35.1733	33.7010	23.1767	15.1380	25.1348	7.7175	19.1719	32.6735	11.0887	25.0392
14	34.9889	33.4009	22.8959	14.7045	24.7310	7.3364	19.0216	32.5856	11.1545	25.0684
15	34.8093	33.1042	22.6206	14.2720	24.3288	6.9635	18.8809	32.5037	11.2379	25.1056
16	34.6347	32.8110	22.3509	13.8406	23.9283	6.6002	18.7500	32.4278	11.3385	25.1508
17	34.4651	32.5214	22.0871	13.4104	23.5296	6.2482	18.6290	32.3580	11.4560	25.2040
18	34.3005	32.2354	21.8294	12.9816	23.1327	5.9095	18.5182	32.2943	11.5898	25.2650
19	34.1410	31.9532	21.5780	12.5543	22.7379	5.5866	18.4177	32.2368	11.7393	25.3340
20	33.9868	31.6750	21.3331	12.1286	22.3451	5.2823	18.3276	32.1854	11.9039	25.4107
21	33.8378	31.4006	21.0950	11.7047	21.9545	5	18.2483	32.1403	12.0830	25.4951
22	33.6943	31.1304	20.8639	11.2828	21.5662	4.7437	18.1797	32.1014	12.2761	25.5872
23	33.5561	30.8644	20.6400	10.8632	21.1804	4.5177	18.1221	32.0688	12.4824	25.6868
24	33.4234	30.6027	20.4236	10.4462	20.7972	4.3270	18.0755	32.0425	12.7013	25.7938
25	33.2962	30.3453	20.2148	10.0319	20.4167	4.1761	18.0400	32.0225	12.9321	25.9083
26	33.1747	30.0926	20.0141	9.6209	20.0390	4.0697	18.0156	32.0088	13.1743	26.0300
27	33.0589	29.8444	19.8215	9.2136	19.6644	4.0112	18.0025	32.0014	13.4272	26.1589
28	32.9488	29.6011	19.6373	8.8104	19.2931	4.0028	18.0006	32.0004	13.6902	26.2949
29	32.8445	29.3626	19.4618	8.4119	18.9251	4.0448	18.0100	32.0056	13.9628	26.4379

(a) Section of section to the distance of each tube

	A	B
	x	y
	Number	Number
1	-18	-16
2	-18	-15.5500
3	-18	-15.1000
4	-18	-14.6500
5	-18	-14.2000
6	-18	-13.7500
7	-18	-13.3000
8	-18	-12.8500
9	-18	-12.4000
10	-18	-11.9500
11	-18	-11.5000
12	-18	-11.0500
13	-18	-10.6000
14	-18	-10.1500
15	-18	-9.7000
16	-18	-9.2500
17	-18	-8.8000
18	-18	-8.3500
19	-18	-7.9000
20	-18	-7.4500
21	-18	-7
22	-18	-6.5500
23	-18	-6.1000
24	-18	-5.6500
25	-18	-5.2000
26	-18	-4.7500
27	-18	-4.3000
28	-18	-3.8500
29	-18	-3.4000

(b) Coordinates (x, y)

Fig. 6.53 Results

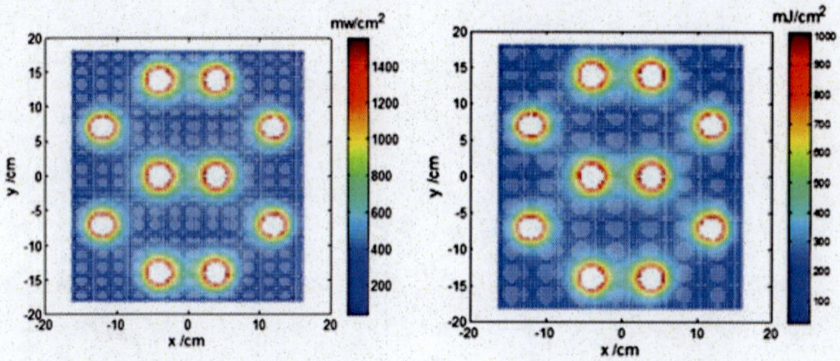

(a) Light intensity (b) Dose

Fig. 6.54 2D distribution of dose and intensity of arbitrary section

As can be seen from the figure, most of the doses accepted by the microorganisms are between 248.1 and 473.8 mJ/cm^2. According to the simulation results, the average UV dose in the cavity is 278.5 mJ/cm^2.

By comparing the simulation results of FLUENT and MATLAB, it can be seenthat the maximum ultraviolet irradiation intensity from FLUENT is 14065.78 W/m^2, and the range of the microbiological dose is between 200 and 400 mJ/cm^2. MATLAB simulated that the maximum UV irradiation intensity was 15,401 W/m^2, and most of the microbially accepted dose values ranged from 248.1 to 473.8 mJ/cm^2. Although

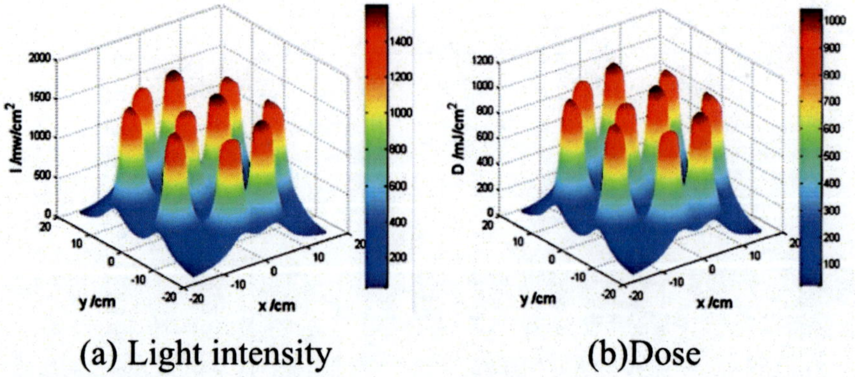

Fig. 6.55 3D distribution of dose and intensity of arbitrary section in cavity

Table. 6.9 Simulation result

Dose /mJ·cm^2		Radiation intensity /mW·cm^{-2}	
Maximum radiation dose	944.1	Maximum radiation intensity	1540.1
Minimum radiation dose	21	Minimum radiation intensity	34.3
Average radiation dose	278.5	Average radiation intensity	455.9

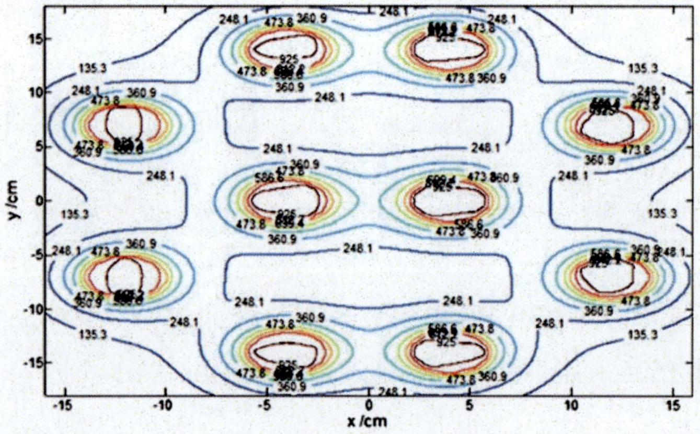

Fig. 6.56 Isoline distribution of radiation dose in uvc

the simulation results of the two methods are different, the overall difference is not significant. Both simulation methods have their own advantages:

① The FLUENT discrete calculation method considers the influence of the hydraulic condition comprehensively, which makes the simulation result more in line with the actual situation. However, the setting of parameters and the division of the lamp grid in the simulation process are complicated.

② The MATLAB method calculates the average dose in the sterilizer based on the product of the average UV irradiation intensity and the residence time. It can directly use the calculation formula of the radiation intensity and dose to program and simulate the distribution of the upper and the lower doses in the UVC sterilizer. Simulation is simpler and more intuitive, and the amount of calculation is small. However, because it does not consider the impact of the sterilizer hydraulic conditions, it will often overestimate the actual UVC dose value of the sterilizer.In addition, the developed MATLABprogram can also be applied to the intensity simulation of other structures of the ultraviolet device, with changes to only some parameters, which means thatit can be used widely.

Both of the methods above can estimate the irradiation intensity and dose value in the sterilizer. In the future research, theycan be used appropriately, according to the actual situation, which can not only greatly reduce the workload of researchers, but also provideuseful methods to improve the structural design.

6.2.2 Optimization of Inert Gas Method and Pharmacy Method Sterilization System

The optimization of sterilization system based on the inert gas method and the pharmacy method is mainly to consider the uniformity of diffusion of gases and agents in ballast water and the speed of diffusion. In general, the electrolysis method and the pharmacy method use the sodium hypochlorite solution to kill microorganisms because of the sterilization mechanism. Therefore, these two methods can be collectively referred to as the pharmacy method.

1. **Inert gas method**

Whether the inert gas is nitrogen produced by molecular sieve filtration or by burning the oxygen in the air via a burner, the gases formed after washing have the effect of inhibiting the growth of microorganisms in the ballast water.

When the vessel is ballasted, the system injects inert gas into the ballast pipeline. In order to enable the inert gas to rapidly dissolve and diffuse in the ballast water and dispel the dissolved oxygen in the effluent, the system uses inert water with back pressure to inject inert gas. Inert gas injection nozzle should be opposite to the direction of flow, as shown in Fig. 6.57. Studies have shown that when inert gas is injected into ballast water with a certain flow pressure, inert gas will quickly diffuse and dissolve in ballast water within a distance of 5 times the diameter of the ballast pipeline into which inert gas is injected.

2. **Pharmacy method.**

Pharmacy law is divided into two direct drug and electrolytic methods. The direct medicament method directly sends the configured sodium hypochlorite aqueous solution to the medicament injection tube by a diaphragm pump. The ballast water passes

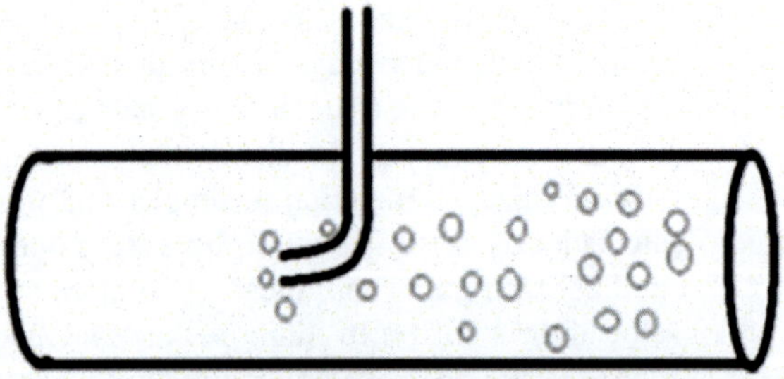

Fig. 6.57 Inert gas injection process

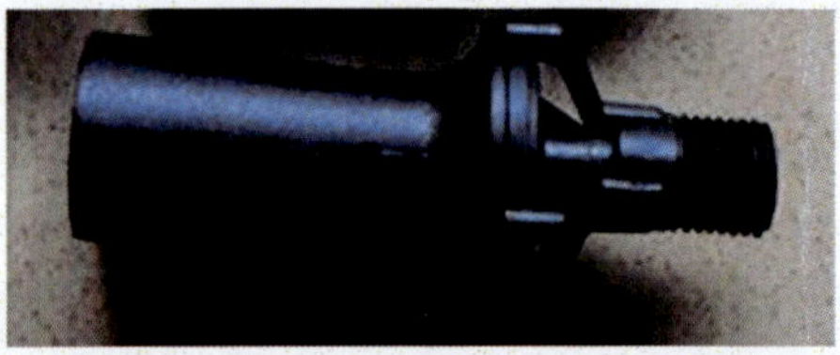

Fig. 6.58 Drainage nozzle contour

through the drainage nozzle, and the injected sodium hypochlorite solution is quickly mixed with the ballast water. The system design of the direct pharmacy method is mainly to optimize the mixing agent and the ballast water's drainage nozzle to avoid the loss of the pressure head of the ballast pipeline. Figure 6.58 shows the appearance of the pharmacy drainage nozzle.

The electrolysis method is intrinsically different from the direct pharmacy method in terms of drug injection. However, since electrolysis is the electrolysis of seawater, sodium hypochlorite solution is directly produced. Therefore, when the vessel enters the freshwater route of an inland river for loading operations, the treatment of water needs to be detected by a salinity sensor, and the working mode of the ballast water system is converted to the freshwater mode in time, and 1 to 2% of split fresh water is introduced from the ballast water, the crude salt is dissolved, and then sent to the tube plate electrolysis. After the electrolysis, sodium hypochlorite is injected back into the ballast water main pipe according to the dose to achieve the effect of killing microorganisms. In order to make the concentration of sodium hypochlorite in the ballast water mixed with the ballast water lower than 10 Ppm that affects the corrosion of the ship ballast tank coating, the TRO sensor that enters the ballast tank piping port will monitor the active material concentration in the ballast water in real time, and control the power of the input electrode of the tube plate type electrolyzer to realize the control of the active material concentration in the ballast water.

References

1. Wang, L., Z. Zhang and Y. Wu et al. 2009. Numerical and experimental study on liquid-solid flow in a hydrocyclone. *Journal of Hydrodynamics, Series B* 21(3): 408–414.

2. Jing, Li. 2012. Numerical Simulation of Sediment Sedimentation in VLCC Ballast Tanks. Dalian:Dalian University of Technology Masters Thesis.
3. Yue-juan, Yan. 2014. Research on Solid-liquid Two-phase Flow Characteristics in Downstream Swirling Sand Remover. Daqing: Northeast Petroleum University, Ph.D. Dissertation.
4. Zhang, J., X.Y. You and Z.G. Niu. 2011. Numerical Simulation of Solid-Fluid Flow in Hydrocyclone. *Chemical and Biochemical Engineering* 25 (1): 37–41.
5. Lei, Feng. 2011. Three-dimensional numerical study of solid-liquid separation cyclone in flue gas desulfurization system. Baoding: Master thesis of North China Electric Power University.
6. B. Wang, K. W. Chu and A. B. Yu. 2007. Numerical study of particlefluid flow in a hydrocyclone. *Industrial and Engineering Chemistry Research* 46: 4695–4705.
7. Li, Luo. 2012. Numerical calculation and performance analysis of solid-liquid two-phase flow in hydrocyclone. Guangzhou:South China University of Technology.
8. B. Wang, and A. B. Yu. 2008. Numberical study of the gas-liquid-solid flow in hdrocyclone with different configuration of vortex finder. *Chemical Engineering Journal* 135: 33–42.
9. Rui, Ying, Wang Wei-bing and Qiu Yan-jun et al. 2015. Design and optimization of hydrocyclone for purifi cation of carbide slag. *Mining Machine* 43 (4): 95–98.
10. Yu-ping, Cao and Jiang Lin-tian. 2015. Research status and development trend on hydrocyclone. *Industrial Water Treatment* 35(2): 11–14.
11. Wei-an, Gong. 1992. Volute design theory and calculation of hydrocyclone. *Petroleum Machinery* 20(12): 1–6.
12. Gui-dong, Yan, Zhang Jin-jun, and Wang Ren-qian. 1990. *Piezoelectric transducer and transducer array (I)*, 231–248. Beijing: Peking University Press.
13. Cheng-yan, Wang, Liu Xi-min and Ran Xiang-cheng et al. 2009. Preparation of nanosized TiO_2 and its applied in COD detection. *China Environmental Testing* 25(3): 66–6. Beijing.
14. JFE BallastAce®. 2013. Ship Ballast Water Management System Operation and Maintenance Manual.Japan: JFE Corporation.

Chapter 7
Development of Control System for Ballast Water Management System

The ship ballast water management system is a regulatory product developed to meet the environmental protection requirements. People use various principles and methods to develop and perfect multi-type products. Based on the requirements of safety, energy saving, environmental protection, convenient operation, automation, and system operation traceability, the company designs and develops control and management systems for its work processes.

Around the ship's ballast and unloading, as well as emergency operations involving ship safety, the main functions of the system control unit in accordance with the requirements of the IMO Convention include ballast water loading management, information recording, system function patrol detection, fault alarm, and automatic operation execution, along with system work process display and so on. For the setting of the control unit of the ballast water management system, in addition to installing the automatic control station in the cabin, it is also necessary to access the centralized remote control station.

Various types of ballast water and sea-keeping biological management systems, due to their different principles and methods, have different modes of operation and automatic control operations. However, its main working process is to store and print information on ballast water loading flow, loading position GPS coordinates, and maneuvering control process; to detect each unit in the system work flow to ensure that the system can effectively kill microorganisms and protect the ship to ensure thatthe basin waters of the route are not attacked by exotic species.

Therefore, the control units of various types of ballast water management systems have installed a variety of sensors to ensure the normal operation of the system according to their respective working principles. In the design of the control unit, industrial controllers are used as much as possible to ensure efficient and reliable operation of the control unit. The touch screen and industrial configuration software are used on the man–machine interface to make the ballast water management and control process straightforward and easy to use.

N. Chen et al., *Development and Implementation of Ship BWMS*,
https://doi.org/10.1007/978-981-33-6865-1_7

7.1 Test Sensors for Ballast Water Management Systems

Due to the different development principles of the ballast water management system, its system detection sensors are not the same. According to the module of the ballast water management system listed in this book, except for the flow sensor that detects changes in the flow rate of the system during the whole process of the system work and the differential pressure transmitter that detects the pressure difference of the pretreatment unit for filtration and filtration, the sterilization is performed. The main types of units include ultraviolet, photocatalytic, inert gases, electrolysis, and chemical agents. Therefore, in addition to the conventional temperature, pressure, and limits, the sensors installed on these sterilization units developed with different principles also have UV intensity, TRO, salinity sensors, and metering pumps [1].

1. **Flowmeter**
   In the ballast water management system, due to the large flow of ballast water delivered, the flow sensor selects electromagnetic flowmeters to monitor the flow variation of the system's ballast and unloading. This is a flow monitoring sensor developed using Faraday's law of electromagnetic induction. It consists of two parts: the flow sensor and the converter. The measuring tube is equipped with an excitation coil above and below, a magnetic field is generated bythe excitation current through the measuring tube, and a pair of electrodes are installed on the inner wall of the measuring tube to make the contact with the liquid to draw the induced potential and send it to the converter. Excitation current is provided by the converter. The advantage of this sensor is that the measurement channel is a smooth, straight pipe with no obstructive flow detection device. It is not easy to be blocked and is suitable for measuring liquid–solid two-phase fluids containing solid particles or fibers; no pressure loss due to detection flow is generated. The resistance is only the resistance along the same length of the pipeline, and the energy saving effect is significant. It is most suitable for the large diameter water supply pipeline that requires low resistance loss. The measured volumetric flow rate is not affected by changes in fluid density, viscosity, temperature, pressure, and conductivity; the straightforward pipe section requires low, large measuring range, typically 20:1–50:1, with a selection of wideflowrange; full-scale liquid flow rate can be selected within 0.5–10 m/s; diameters can range from a few millimeters to 3 m. It can measurepositive and negative two-way flow, can also measure pulsating flow, as long as the pulsation frequency is less than the excitation frequency; instrument output is essentially linear; the type of material in contact with the fluid can be found easily; and it can be applied to corrosive fluids.
   The shape of the flowmeter is shown in Fig. 7.1. Its main parameters are shown in Table 7.1.
   The installation in the ballast water management system is shown in Fig. 7.2. As can be seen in the right figure, the correct installation position of the flowmeter should be selected in the straight section full of liquid, such as the

Fig. 7.1 Electromagnetic flowmeter

Table 7.1 Main parameters of flowmeter

Load resistance	0–1.5KΩ
Frequency output range	1–5000 Hz
Output pulse range	0–100 pulse/s
Highest withstand voltage	36 V, DC
Maximum load current	250 mA

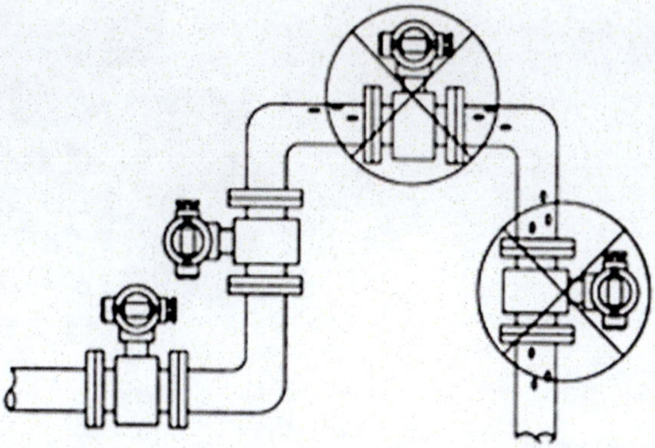

Fig. 7.2 Installation of flowmeter in BWMS

vertical section of the pipeline (flow direction from bottom to top is appropriate), or a horizontal pipeline full of liquid (appropriately at the lowest point of the whole pipeline), and no partial pipeline condition may occur during the measurement.

2. **Differential pressure transmitter**
   The differential pressure transmitter in the backwash filter unit monitors the differential pressure on both sides of the backwash filter screen in real time, controls the start and stop of the filter washer motor, and cleans the filter screen, thereby improving the flowability of the filter screen and preventing the filter from being clogged with impurities. The appearance of the differential pressure transmitter is shown in Fig. 7.3. Its main parameters are shown in Table 7.2. The pressure transmitter directly measures the pressure acting on the diaphragm surface. When the pressure causes the diaphragm to deform slightly, the high-precision circuit on the diaphragm will transform this tiny deformation into a highly linear pressure proportional to the pressure. The voltage signal is also proportional to the excitation voltage, and then it uses a dedicated chip to convert

Fig. 7.3 Pressure difference transmitter

Table 7.2 Main parameters of pressure difference transmitter

Power supply	24 V, DC
Operating temperature	−40~+ 80 °C
Protection level	IP 65
Materials	Aluminium alloy

this voltage signal to an industry-standard 4–20 mA current signal or 1–5 V voltage signal.

3. **Temperature sensor**
The temperature sensor is mainly used to monitor the temperature of the ultraviolet and photocatalytic sterilizing chamber, to prevent the working water temperature from being excessively high to cause fouling on the quartz sleeve of the ultraviolet lamp tube or the high temperature in the chamber caused by the poor fluidity of the seawater in the ballast water pipeline damages the sealed original. The shape of the temperature sensor is shown in Fig. 7.4. The main technical parameters are shown in Table 7.3.
There are many types of temperature sensors, such as thermocouples, thermal resistanceetc. Since the temperature of the ultraviolet sterilizing unit chamber is preset to 50 °C, the thermal resistance is selected for the temperature sensor. Usually, the thermal resistance is composed of a resistor body, an insulating sleeve, and a junction box. It is made of a copper wire or a molybdenum wire wound on an insulating skeleton and then inserted into a sheath. The production process uses double lines. As forwinding, its purpose is to make the thermal

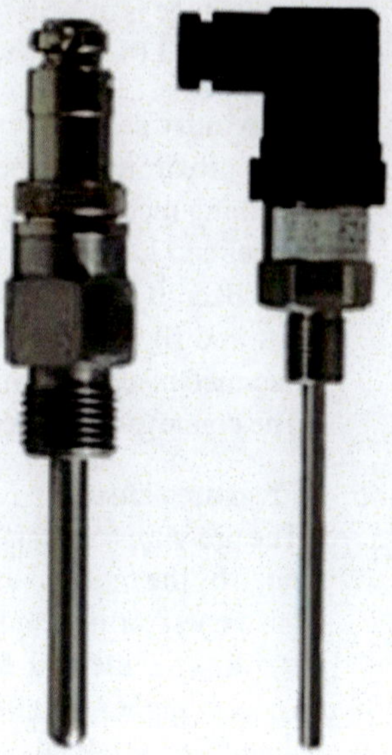

Fig. 7.4 Temperature sensor

Table 7.3 Temperature sensor parameter

Specification	0–150 °C
Power supply	24 V, DC
Output	4–20 mA
Accuracy	0.5%
Mounting thread	M20 × 1.5

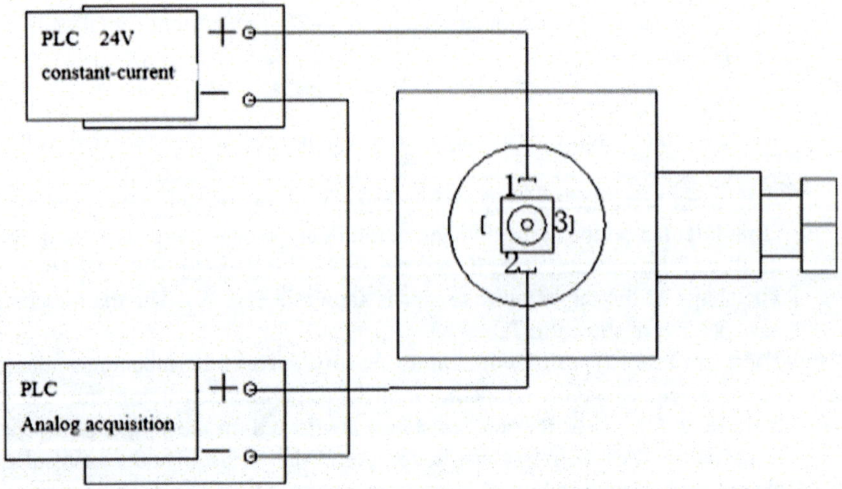

Fig. 7.5 Temperature sensor and PLC wiring method

resistance without additional inductance. Copper thermal resistance measurement range is of −50 °C~+150 °C in temperature. The wiring of the temperature sensor and PLC is 4-wire, and the wiring method is shown in Fig. 7.5.

4. **Proximity sensor**
   The proximity sensor is mainly used to monitor the movement limit of the cleaning mechanism on the filter and the UV sterilizer. It is used to monitor the upper and lower positions of the cleaning nozzle moving up and down. For the sterilizer, it is used to monitor the UV lamp. The quartz jacket scraper moves left and right. The proximity sensor is shown in Fig. 7.6. Its main technical parameters are shown in Table 7.4.
   The connection between the proximity sensor and the PLC is shown in Fig. 7.7.

5. **Pressure sensor**
   The pressure sensor is mainly used to monitor the pressure of nitrogen gas separated by the molecular sieve or the pressure of the inert gas generated by the inert gas generator injected into the ballast water by the inert gas oxygen-removing sterilization method. Ensure that the partial pressure ratio of the injected ballast water is higher than that of oxygen dissolved in the ballast seawater to drive off the oxygen in the ballast water to achieve the sterilization effect. The shape of the pressure sensor is shown in Fig. 7.8. Its parameters are shown in Table 7.5. The working principle of the pressure sensor is that the pressure acts on the wafer through the diaphragm or piston to generate the charge. After being amplified, the display or recording device displays or records the pressure to achieve the pressure measurement. It is characterized by wide frequency response, large measurable pressure range, small size, light weight, easy installation, and multi-directional pressure measurement.

Fig. 7.6 Proximity sensor

Table 7.4 Proximity sensor parameter

Operating temperature	−25~+75 °C
Voltage range	10−30 V, DC
Current consumption	< 10 mA
Load current	< 150 mA
Reverse connection protection	Have
Protection level	IP67

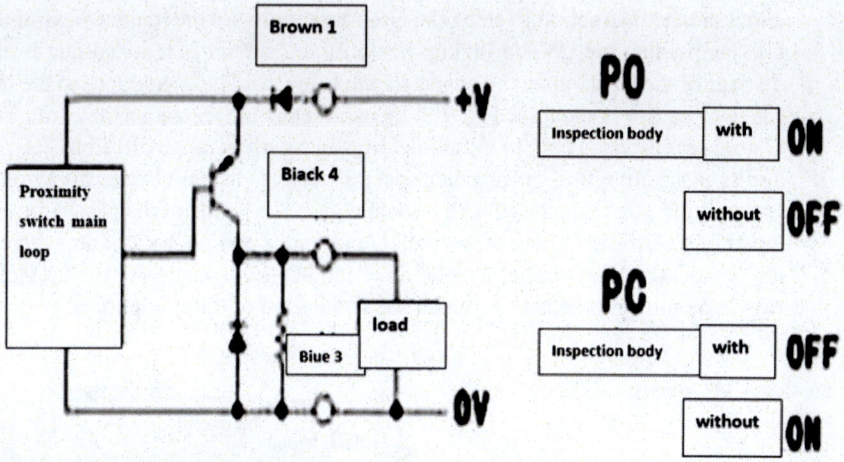

Fig. 7.7 Proximity sensor and PLC's wiring

Fig. 7.8 Pressure sensor

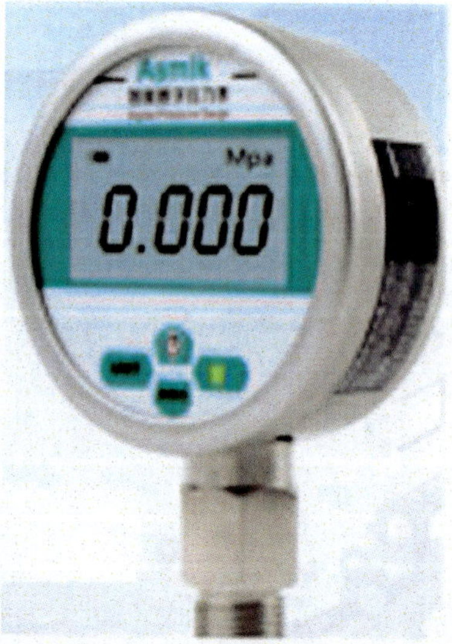

Table 7.5 Pressure sensor parameter

Specification	0– 10 kPa
Power supply	24 V, DC
Output	4–20 mA
Accuracy	0.5%
Mounting thread	M20 × 1.5

6. **UV intensity sensor**
The UV intensity sensor is used to detect the UV intensity in the sterilizer cavity, and issue the instruction to clean the lamp jacket or stop the replacement of the UV lamp when the UV sterilizing dose in the chamber is reduced due to the damage of the UV lamp or the quartz sheath is fouled. The appearance of the UV intensity sensor is shown in Fig. 7.9. Its parameters are shown in Table 7.6. The UV sensor responds only to ultraviolet rays in a narrow range of 185 to 260 nm and is insensitive to light in other spectral ranges. It is a closed, ultraviolet-transmitting glass tube filled with a special gas. The inside of the glass tube has a pair of electrodes led out of the metal lead-anode and photocathode, wherein the photocathode is made of a metal material that is only sensitive to ultraviolet rays and emits photoelectrons under the irradiation of ultraviolet rays.

Fig. 7.9 UVC intensity sensor

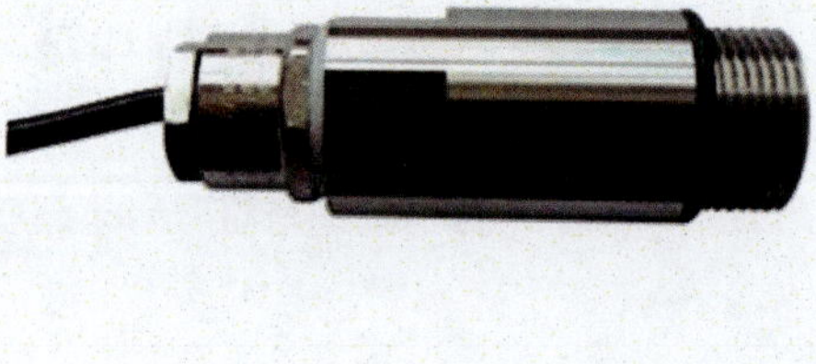

Table 7.6 UVC intensity sensor parameters

Output signal	4–20 mA linear output
Power supply	24 V, DC
Load impedance	$250 - 680\Omega$
Power consumption	< 0.5 W
Operating temperature	−10~60 °C
Connection method	G3/4′ External thread
Material	Stainless steel

7. **TRO sensor** [2]
The TRO sensor, also known as the Total Residual Oxidized Material Detector or the Residual Chlorine Detector, is an important parameter for judging whether the active ingredient in water is sufficient. The sensor is mainly used in the bactericidal ballast water system of the pharmacy method and electrolysis method. It is used to monitor the concentration of residual oxides in the ballast water when ballast water is unloaded to help the system confirm whether to add neutralizer to the unloaded ballast water. The shape of the TRO sensor is shown in Fig. 7.10. Its parameters are shown in Table 7.7.
The TRO sensor uses the cuvette method to measure the amount of chlorine in water. It can automatically measure and display readings with high reliability and measurement accuracy.

Fig. 7.10 10TRO sensor

Table 7.7 TRO sensor parameters

Measuring range	0.00–15.00 mg/L (PPM)
Accuracy	± 5%
Resolution	0.01 mg/L (PPM)
Communication port	RS-485 port with Modbus RTU protocol
Operating temperature	0–55 °C(32–131 °F)
power supply	AC100/220 V, 50/60 Hz or DC 24 V
Environmental conditions	≤ 95% Relative humidity (non-condensing)

8. **Salimeter** [2]
The salimeter is a monitoring instrument for determining whether a ship is in an inland river or in a marine navigation by the electrolytic sterilization method, so that the electrolytic ballast water treatment system can automatically determine whether it is necessary to switch to an operation mode in which sodium hypochlorite is produced by using a coarse salt to prepare an electrolyte. The salimeter can accurately measure conductance, salinity, TDS, and temperature. The built-in microprocessor calculates and compensates for the measured parameters. Using DC24V power supply, the color LCD display can display TDS, salinity, conductance and temperature in real time and output measurement signals in a variety of ways. Sensors use inductive probes that are electrodeless and therefore have no polarization response. It can measure heavy media and sediment-prone media and solutions. The measurement and the medium are completely electrically isolated. The working principle is mainly to use the light to change from one medium to another, and the ratio of the incident angle sine is constant. The ratio is called the refractive index. By using the content of the soluble substance in the salt solution and the refractive index in the normal environment, the refractive index of the salt solution can be determined.Based on that, the salimeter/refractometer can calculate the salt concentration. The appearance of the salimeter is shown in Fig. 7.11. Its parameters are shown in Table 7.8.

9. **Metering pump** [3]
Metering pumps are devices used by pharmacy and electrolysis methods that control the injection of sodium hypochlorite solution or neutralizer solution into the ballast water during ballast or unloading.
Metering pumps have diaphragm type, reciprocating plunger type, gear type, peristaltic type, screw type, etc. These pumps use solvent methods to quantitatively deliver medicaments. Due to the large amount of drug used in the ballast water system, diaphragm pumps are often used in the system. The outline of the diaphragm pump is shown in Fig. 7.12. Its parameters are shown in Table 7.9.
Diaphragm pump features include: adjustable eccentric mechanism, flat pulsation; pump head structure can choose from—PVC, PVDF, 316SS, high viscosity

Fig. 7.11 Online salimeter

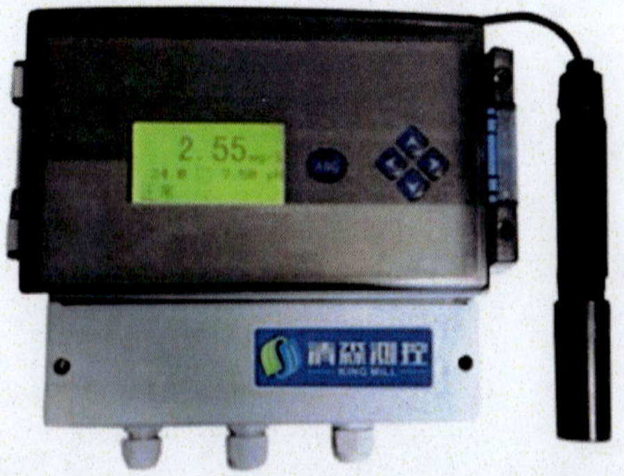

Table 7.8
Salimeter parameters

Measuring range	Concentration: 0.00–20.00% NaCl
Resolution	0.01%, 0.1 °C
Electrode constant	1–55.0 cm is set by software
Warming range	Automatic 0–99.9 °C, Manually 0–99.9 °C
Accuracy	± 2.0% (Calibration error ≤ 0.05%), ± 0.3 °C
Galvanically isolated output	4–20 mA (load < 750 Ω)
Alarm relay	AC220 V, 3A, Alarm signal isolation output
Communication Interface	Isolation RS485 (Optional)
Power supply	AC220 V ± 22 V, 50 Hz ± 1 Hz
Protection level	IP65
Working conditions	Ambient temperature 0–60 °C, humidity ≤ 90c/o
Weight, size	0.8 kg, 200 × 150 × 108 mm;

and slurry pump head; mechanical drive PTFE diaphragm; material side without diaphragm guard plate, easy for material to get hrough; with three automatic control methods for a variety of control needs, manual / electric stroke adjustment.

10. **Electric butterfly valve**
The electric butterfly valve is installed on the pipeline of the ballast water management system. Its role is to control and regulate the seawater entering the system, thereby realizing ballast, unloading, emergency ballast and emergency unloading. The appearance of the electric butterfly valve is shown in Fig. 7.13. Its parameters are shown in Table 7.10.

Fig. 7.12 Diaphragm pump

Table 7.9 Diaphragm pump parameters

Stroke	Adjustable under both static and dynamic conditions
Maximum liquid temperature	40 °C
Steady-state accuracy	± 2%(10–100%)
Maximum suction	3 m water column
Maximum suction pressure	20 m water column
Optional structure	Double diaphragm pump head
Recommended maintenance period	≤ 1.0 MPa 4000 h; ≤ 0.7MPa 8000 h

7.2 Control Unit Design for Ballast Water Management Systems

The control unit of the ballast water management system is the brain of the ballast water loading and unloading management. It is mainly managed by the control module to monitor the power module, detection feedback module, and execution module of the entire system, strictly according to IMO's relevant requirements. The ballast water loading and unloading process is controlled to ensure the normal operation of the system [4].

1. **Analysis of Control Process of Ballast Water Management System**
   As mentioned earlier, ballast water management systems are regulatory products developed to meet the environmental protection requirements. People have developed various types of products with different principles and methods. Therefore, from the perspective of the working process of these systems, it

Fig. 7.13 Electric butterfly valve

Table 7.10 Electric butterfly valve parameters

Rotation angle	0–90° Adjustable
Ambient temperature	−25~+60 °C
Body execution standard	GB/T 3036–94
Power type	AC220V
0–90° opening and closing time	30 s
Electric actuator material	Aluminum alloy
Electric actuator protection class	IP67

is nothing more than the switching control of the valve group during the loading and unloading of ballast water, the cleaning and control of the filter of the pre-filter unit, and the control of the sterilization units developed with different principles or methods.

For example, the ultraviolet or photocatalytic sterilization unit performs the on/off control of the ultraviolet lamp and the inspection of the ultraviolet lamp working status, and controls the cleaning mechanism of the ultraviolet sterilization unit, according to the illumination intensity of the ultraviolet light source and the limit switch of the cleaning mechanism, controls the cleaning mechanism drive motor to forward or reverse; According to the working order of the photocatalytic sterilization unit, the formic acid cleaning pump and the related electromagnetic valve are activated after the ballasting or unloading ends, and the inside of the sterilizer chamber and the lamp tube jacket are descaled and cleaned. The sterilizing unit controls the ultrasonic cleaner on the sterilizing unit and cleans the sterilizer during the ballasting or unloading of the system.

As another example, for the inert gas sterilization system, the operation and management of the system is relatively simple.The system's pre-filtration processing unit uses a membrane filter with higher filtration accuracy, the porosity is only 10 μm, similar to the operation of the backwash filter, when the ballast water passes through the membrane element, the suspended solid impurities in the water are intercepted in the membrane element, indicating that the filtered water is discharged. If the membrane hole is blocked by a large amount of suspended matter, the pressure difference between the inlet and outlet of the membrane module will increase. When the pressure difference's pre-set value is reached, the recoil pump will automatically start. Then the compressed air will be turned on, and the filtered water of the membrane element will become backwash. In the water, the surface of the membrane is subjected to gas–water mixing and flushing under the differential pressure. At the same time, the system opened the nitrogen tank and injected nitrogen into the water through the filtered clean water injection chamber. In order to ensure the continuity of the ballast water injection, the system uses a plurality of membrane filter units to control the pressure difference mode or time mode during back flushing. When the differential pressure controller reaches the system's pre-set value of 0.05 MPa, the differential pressure switch is closed, the recoil pump is started, the backwashing process is executed, and the process is stopped after 30 s, or when the system operation reaches the time mode set for 30 min, the system receives Backflushing instructions, the inlet valve of membrane group 1 is closed for 30 s, the recoil pneumatic is opened for 30 s, the compressed air solenoid valve is opened for 10 s, the recoil pump is started, the membrane group 1 is recoiled, and after 30 s, the next membrane group is inverted Flushing, when the backwash of the *n*th membrane group is completed, the system stops running, and after 30 min, it enters the next cleaning cycle.It can be seen that the control of the ballast water management system is not only the start-stop control of the nitrogen molecular sieve generator, but also the pressure monitoring of the nitrogen gas storage tank, which is mainly the sequential cleaning control of the membrane filtration.

As one more example, the electrolysis method for preparing sodium hypochlorite and the pharmaceutical method is directly added to the sodium hypochlorite sterilization system. In addition to the control system of the pre-filtration treatment unit, the control system performs timing control to prevent it from blocking. The main control is the control of the dosing system and the electrolysis system. In other words, the main object of control is ballast loading. In addition to starting the normal ballast piping valve, the electrolysis water branch bypass electromagnetic valve must be opened to guide1 to 2% of ballast seawater into the tube plate type electrolytic cell, along with the start of DC electrolysis power supply, opening of the dehydrogenation fan and the injection of the metering pump, injection of sodium hypochlorite solution into the ballast water. When the ship is unloaded, it should be based on whether the active ingredient content in the unloading water monitored by the TRO sensor is greater than 0.1 ppm. If it is greater than 0.1 ppm, it opens the neutralization system solenoid valve and starts the neutralization agent delivery metering pump to inject the unloaded ballast water, avoiding to cause active substance pollution in unloaded sea areas. For the pharmacy method system, the control and management of itscontrol system is more convenient. In addition to the automatic opening of the control system valve during loading or unloading, it is the time sequence cleaning of the pre-filter unit or the set time for the ballast. The process starts the metering pump and injects the agent into the ballast water according to the metering. Duringunloading it is based on the active substance content in the unloadedwater monitored by the TRO sensor.When the active material content is greater than 0.1 ppm, the solenoid valve and neutralization agent delivery of the neutralization system are opened. The metering pump injects a neutralizer into the unloaded ballast water.

From the above analysis, the ballast water management system controls the ballast water management system in accordance with the regulations of the IMO. The control and management method includes the following steps: (1) Press and carry water loading and unloading, to control the valve blockof the ballast water management system; (2) Operation and management of the pressure difference and sequence program of the cleaning mechanism of the ballast water pretreatment filter unit; (3) regulation and management of the power of the sterilization unit and running inspection; (4) thecleaning ofthe sterilizer during or after the work. This book will use the UVC sterilization system as an example to emphasize the design technology of the ballast water management system control unit.

2. **Ballast water control unit design** [4, 5]

From the perspective ofthe reliability of the system operation, the manufacturers of current control system of the ballast water management system choose the industrial control PLC module to build its control unit, combined with the industrial touch screen and the use of configuration software developed for man–machine interfacing, forming a complete control unit module that satisfies IMO's specifications for ballast water management. PLC as the core component of ballast water management system control unit has: (1) Anti-interference and high reliability; (2) modular structure, flexible and convenient to use; (3) simple programming and easy to popularize; (4) can be modified online; (5) the network communication function facilitates the decentralized measurement and control system; (6) compared with traditional control methods, the advantages of simple circuits. The PLC omits the system developer's design of the circuit, only through the selection of the PLC module and the wiring according to the control signal, and the development of the control program according to the logic requirements of the management, thus making the development of the ballast water control unit convenient and easy, as shown in Fig. 7.14 for the outline of the PLC. It can be seen from the figure that when designing the PLC controller, in addition to selecting the PLC master module PLC1, the designer also selects PLC2, PLC3 and so on, according to the number of control points of

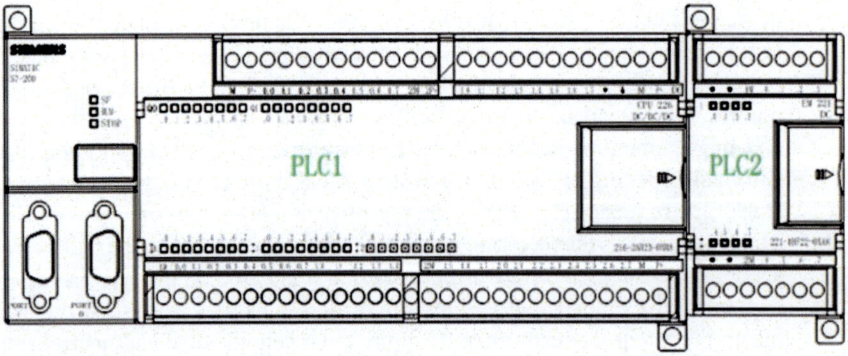

Fig. 7.14 Shape diagram of PLC

the control unit and the type expansion of the control point input/output signals, forming thedigital input/output control and analog input/output control.

As can be seen from the figure, the PLC1 main control module is a monolithic structure, which contains the power supply, CPU, memory, and I/O of the programmable logic controller. The integration of the entire control system revolves around this basic unit structure. It includes: (1) MPI communication interface; (2) DP interface; (3) memory card socket; (4) DO top power supply and output terminal; (5) DI bottom power supply and input terminal; (6) RUN/STOP switch potentiometer expansion I/O connection; (7) status LED.

Taking a 500 m^3/h UVC ballast water treatment system as an example, the development of its control system includes two cabinets, a control part and a power supply part. The control part mainly sterilizes each valve, cleaning motor, and UVC of the ballast management system. The on–off management of the lamp rectifier, after the system starts, completes the switch start-up adjustment of the ballast or discharge valve through the sequence program, completes the power on and off management of the UVC germicidal lamp rectifier, and the output is a switching signal in the system. During operation, the analog signal such as differential pressure signal, ultraviolet illuminance signal and temperature signal are monitored, and according to the system's programmed judgment procedure, the pressure difference and ultraviolet illuminance and temperature monitoring threshold are judged, and kickback is initiated. The drive motor and drain valve of the filter or sterilizer cleaning mechanism are used to clean the filter and the sterilizer. The directional control of the cleaning motor is performed through its respective limit switch, and the cleaning is performed according to the sequence of the program. The design of the control section bread and wiring module is shown in Fig. 7.15.

The control schematic of the system is shown in Fig. 7.16. It can be seen from the figure that the AC power sources L01 and L02 respectively supply power to the touch screens of the PLC and the industrial control machine, and at the same time supply power to the DC power supply UR1. UR1 provides 24VDC

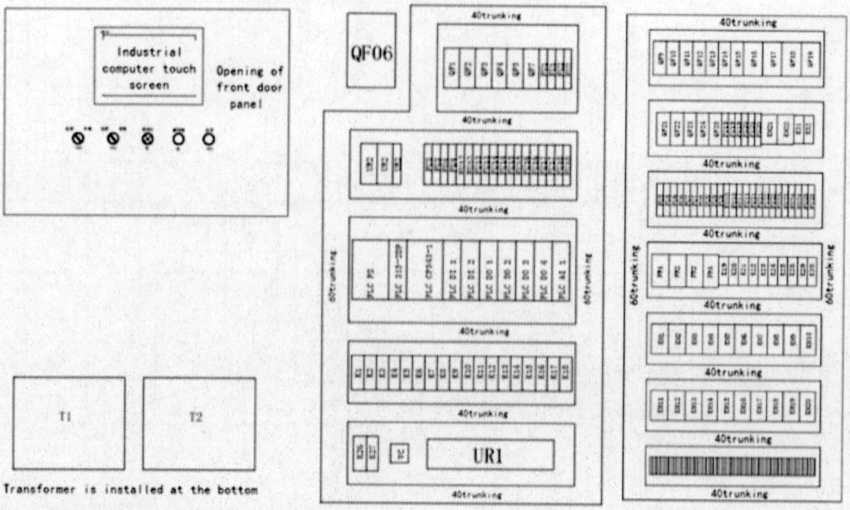

Fig. 7.15 Control panel, positive and negative wiring diagram

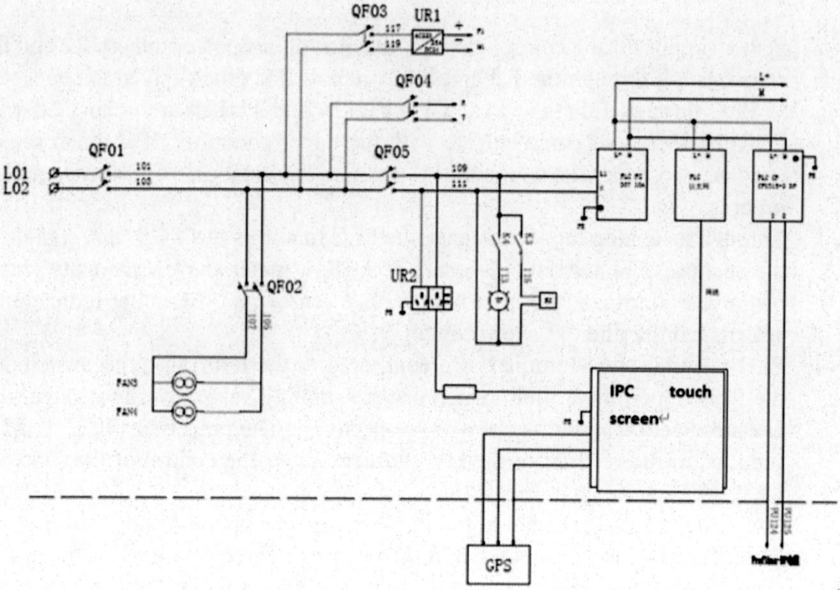

Fig. 7.16 Control system principle diagram

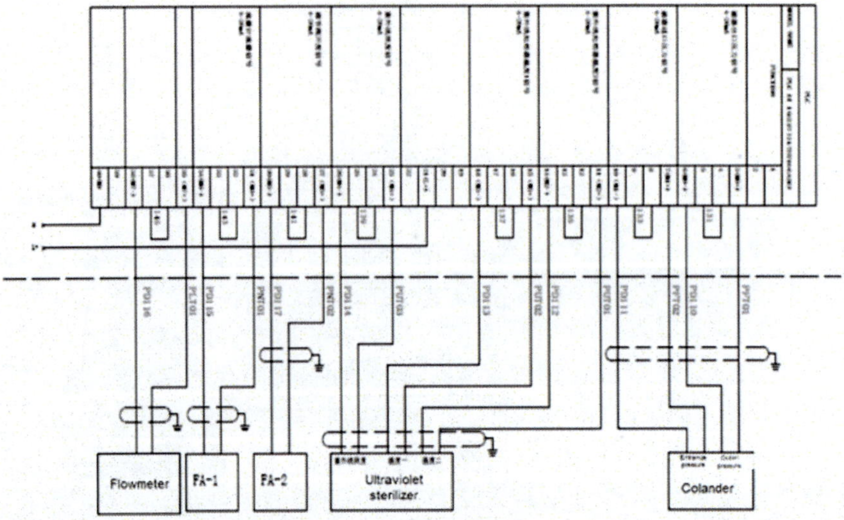

Fig. 7.17 PLC analog input terminal wiring diagram

power supply for the control cabinet; and it supplies power tothe UR2 and the sterilizer YP through the K3 and K4 switches. IPC touch screen is connected to PLC through Ethernet. PLC CP's PU124 and PU125 are connected with PROFIBUS bus to communicate with the upper computer. IPC touch screen is connected with GPS to record the position information of system operation process.

Through its analog input terminals, the PLC monitors the 4–20 mA signals of the inlet pressure and outlet pressure of the flow meter, the UV intensity sensor of the UV sterilizer, the inlet temperature sensor and the outlet temperature sensor, and the filter, as shown in Fig. 7.17.

PLC's digital input terminals are connected to the left/right limit switch and the filter upper/lower limit switch of the sampling valve, overflow valve, and UV sterilizer cleaning mechanism, respectively. The power supply L +, M is supplied to the rectifier of the UV sterilizer under the control of the switches SA1, SA2, as shown in Fig. 7.18.

The switch output terminal of the PLC controls the opening and closing of the butterfly valve No. 1 to No. 6 and the opening degree of the regulating valve according to the sequence program of the PLC preset according to the working procedures of the ballast and the unloading, and controls the ballast pumps No. 1 to No. 2 according to the measurement variation of the flow meter. The operation of the ballast pump is shown in Fig. 7.19.

Figures 7.20 and 7.21 respectively show the start-up and power-up control of the rectifiers of UV lamps No. 1 to No. 12 at the PLC output terminals and the connection between the rectifier and the UV lamp.

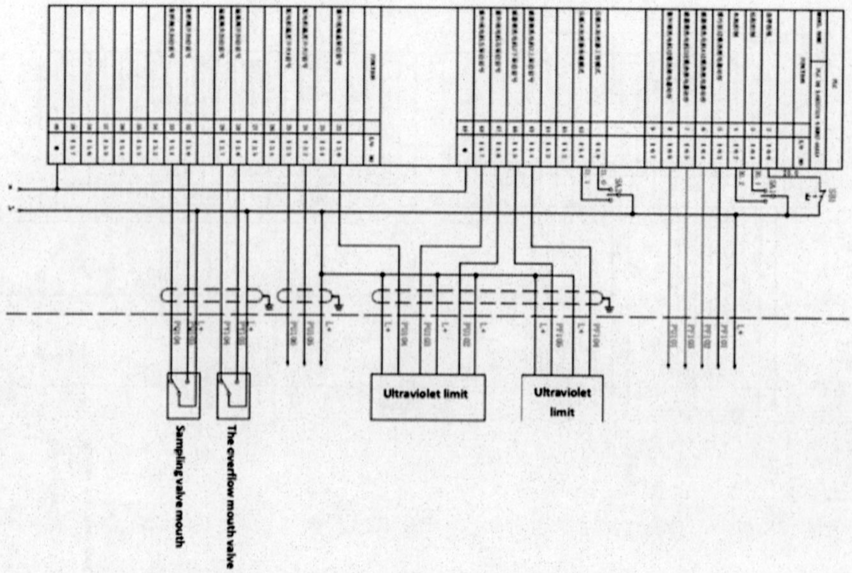

Fig. 7.18 PLC switch input terminal wiring diagram

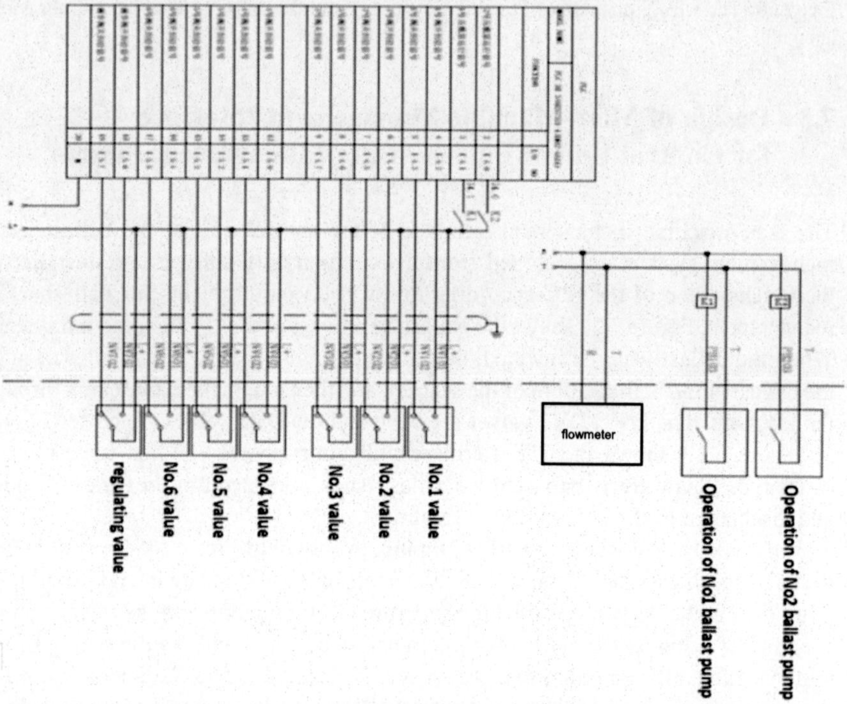

Fig. 7.19 PLC digital output terminal control butterfly valve and ballast pump wiring diagram

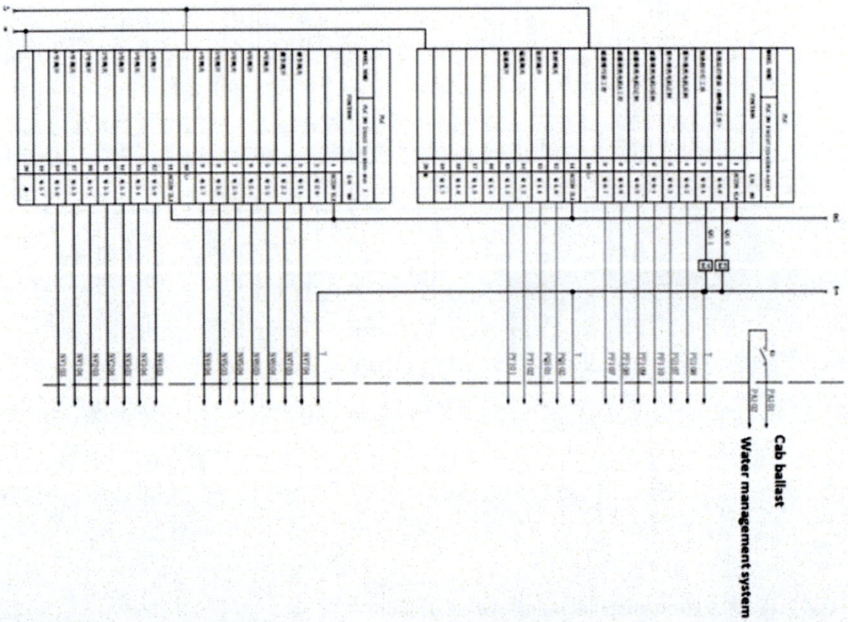

Fig. 7.20 The wiring diagram of PLC digital output terminal and 1–12 rectifier and ultraviolet lamp tube

7.3 Design of Man–Machine Management Interface for Control Unit of Ballast Water Management System

The man–machine management interface of the control unit of the ballast water management system is developed from the configuration software pre-installed in the touch screen of the IPC according to the IMO requirements for ballast water management. Figure 7.22 shows the ballasting process of the UV disinfection system. When the ballast water is injected, the seawater enters the backwash filter through the ballast pump at the bottom of the seaport, and then enters the ballast tank through the UV Unit. The DV1, DV2 valves are open and the DV3 valves are closed [4, 6].

Figure 7.23 shows the UV sterilizer unloading process. When ballast water discharges: Seawater is ballasted by ballast tanks into medium-pressure UV units and discharged to the outboard after treatment.

When the ballast water system is running, as shown in Fig. 7.24, the system first needs to check whether all the valves and meters of the system are in good condition. After that, it checks whether the high-pressure cleaning pump and the sewage pump are stuck.If necessary, it starts the system and once again, checks the wiring of the system. Then, after pressing the system start button, the power indicator lights and the touch screen entersinto the running state. It checksthe system's various devices, meters, and wirings again tomake surethat all components are working normally.

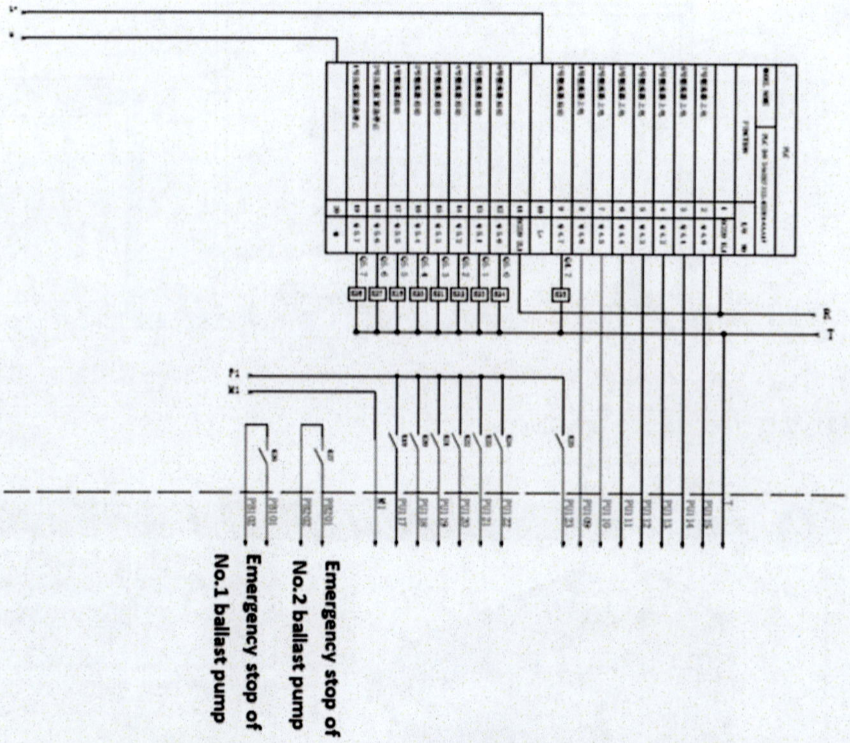

Fig. 7.21 The wiring diagram of PLC digital output terminal and 1–12 rectifier and ultraviolet lamp tube

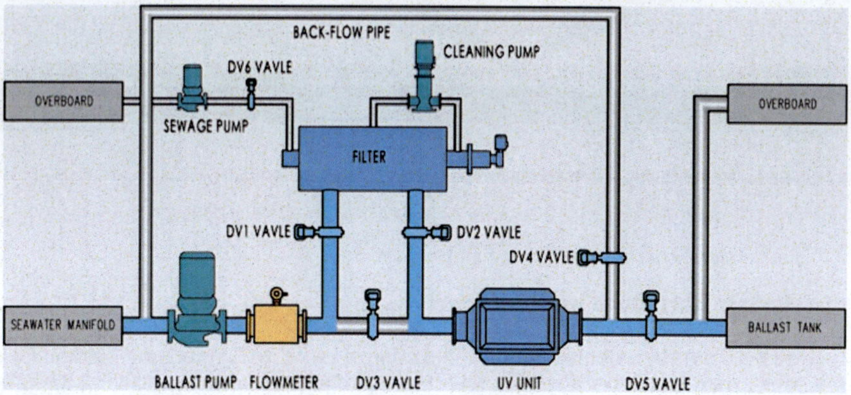

Fig. 7.22 The ballast process of UVC sterilization system

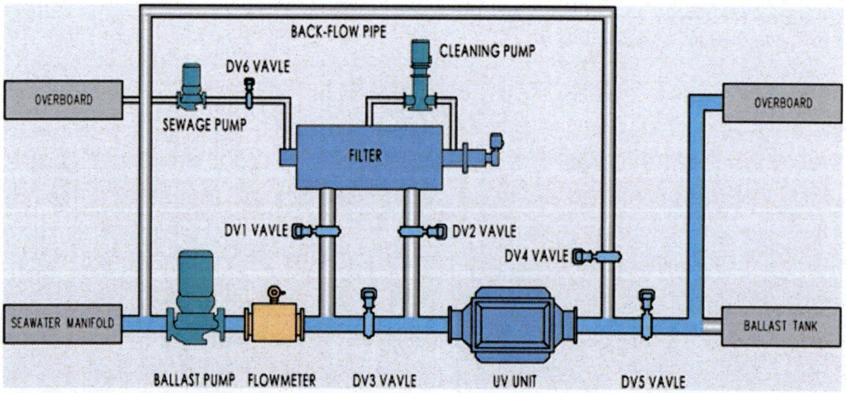

Fig. 7.23 UVC sterilizer unloading flow

Fig. 7.24 Ballast water management system

The system operation steps are as follows:

1 **Ballast start mode working interface**

(1) Turn the selector knob on the control panel to "Automatic" mode. Click the "enter the system main interface" button on the display to enter the main interface of the system, as shown in Fig. 7.25.

(2) Select "left machine", "right machine" or "dual machine" mode according to equipment and ship conditions. Select "Live" or "Remote" under the "Control Location" option. The selected item changes from dark green to light green, as shown in Fig. 7.26.

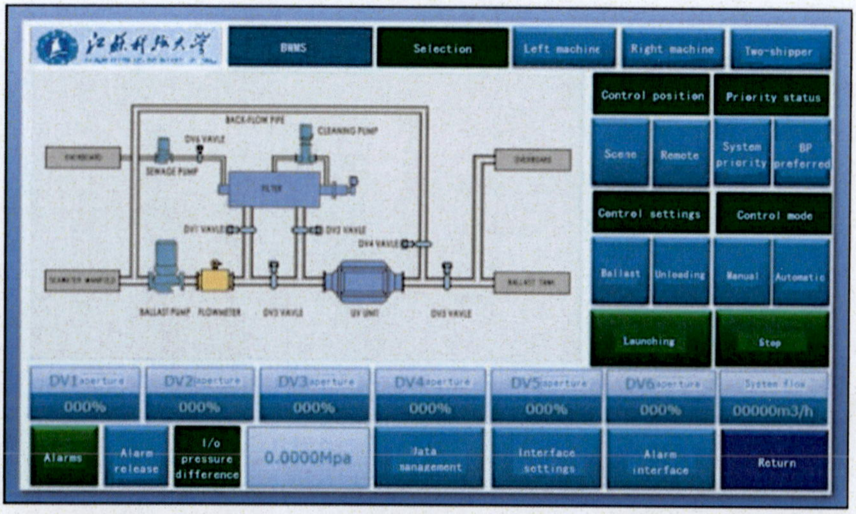

Fig. 7.25 Ballast management System interface

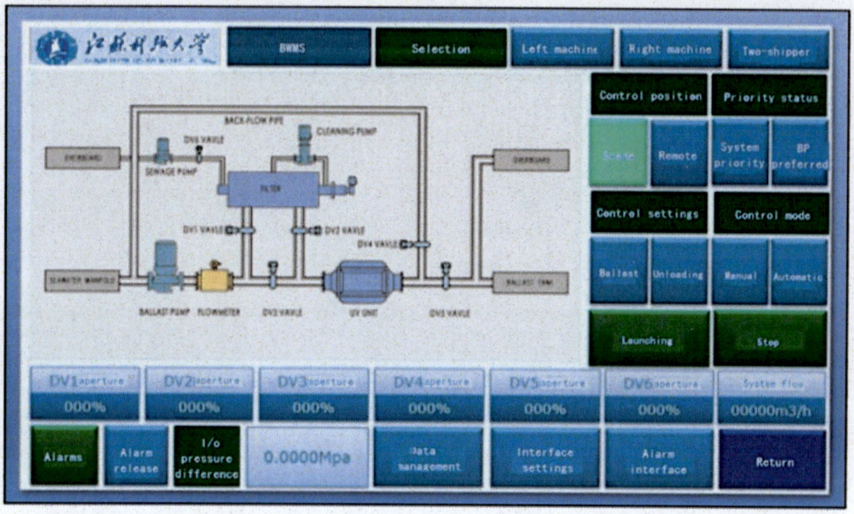

Fig. 7.26 Ballast operating mode selection

(3) Select "System Priority" under the "Priority Status" option. The selected item changes from dark green to light green, as shown in Fig. 7.27.

(4) Select Ballast under the Control Settings option. The selected item changes from dark green to light green, as shown in Fig. 7.28.

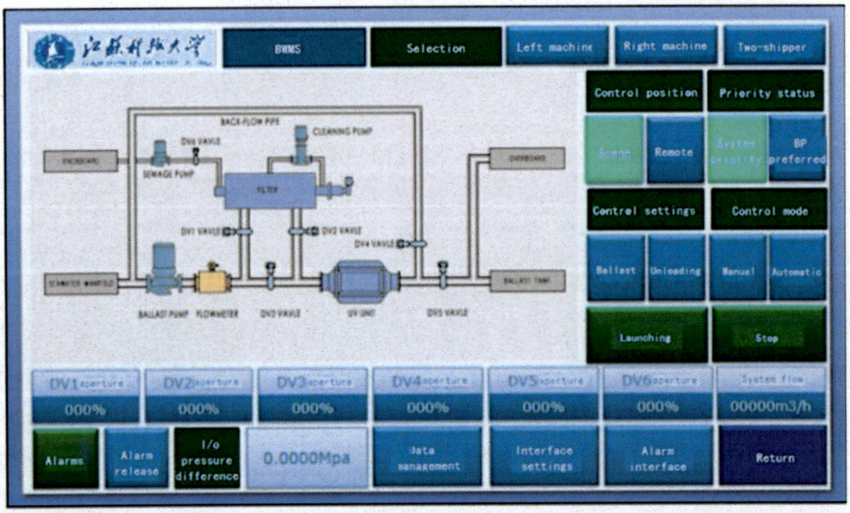

Fig. 7.27 Ballast preferred mode selection

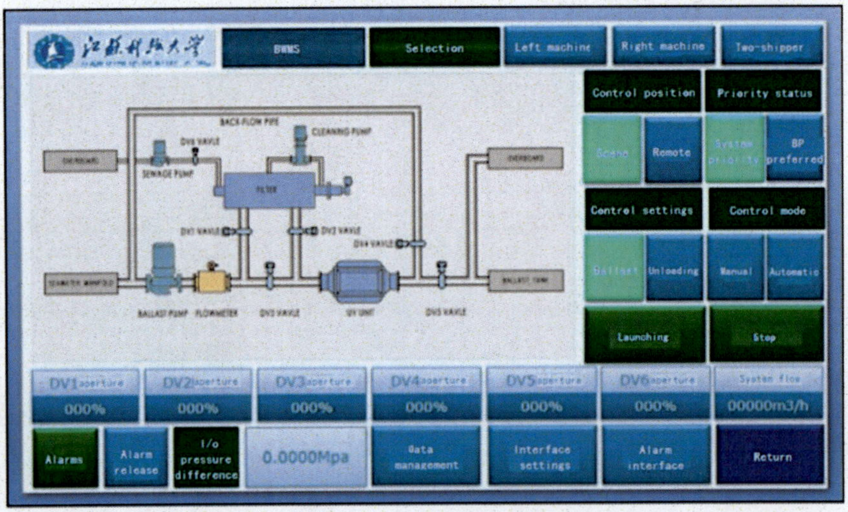

Fig. 7.28 Ballast control method setting

(5) Select “Automatic” under the “Control mode” option. The selected item changes from dark green to light green, as shown in Fig. 7.29.

(6) At this time, the preparation of the system ballast mode control interface is ready. Check the status of each unit and equipment in the system again and make sure it is normal. Click on the system startup option and the system

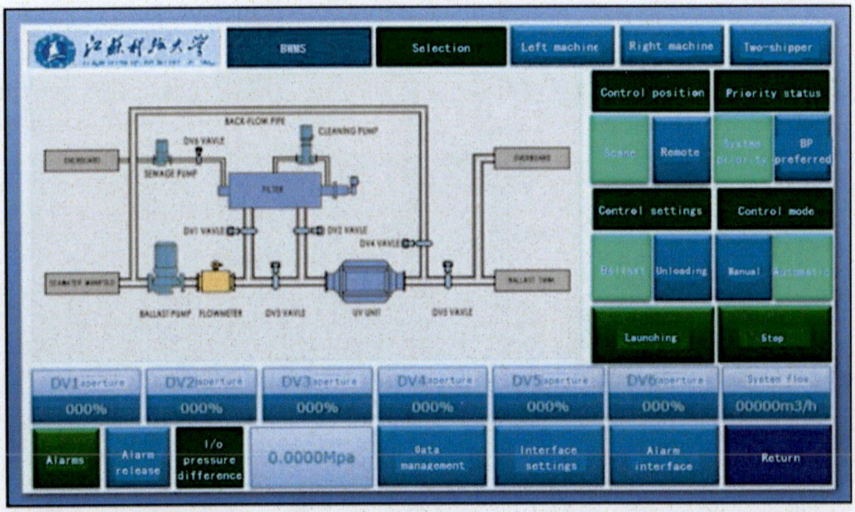

Fig. 7.29 Automatic ballast control method

starts to run in ballast mode. The selected item changes from dark green to light green, as shown in Fig. 7.30.

2. **Ballast start mode working process**

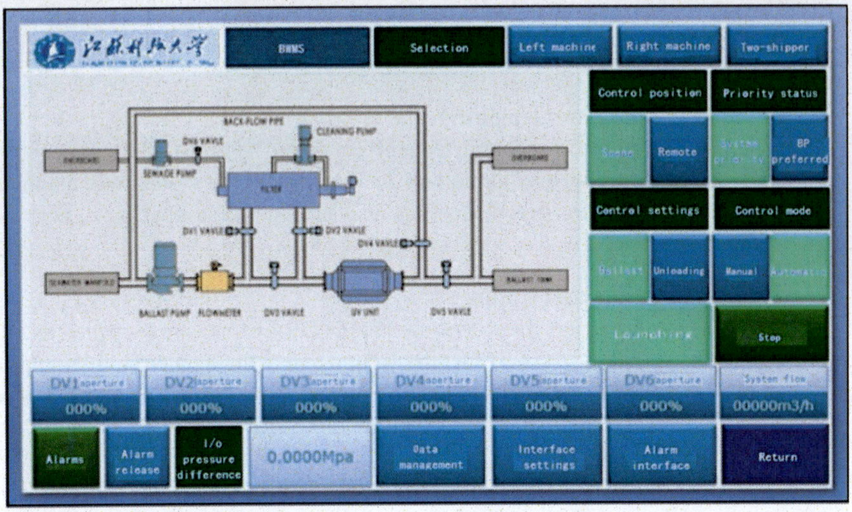

Fig. 7.30 System ballast operation mode

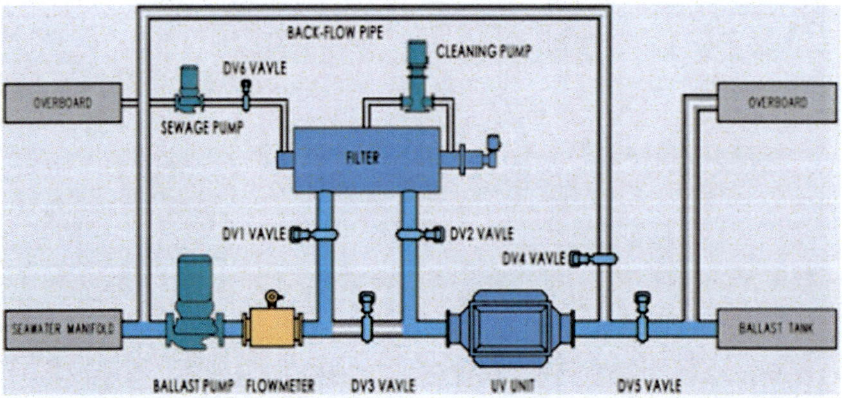

Fig. 7.31 Ballast mode flow chart

After the system is started, the DV1 valve, DV2 valve, and DV4 valve are opened. After the valve is opened, the ballast pump starts. The seawater flows into the backwash filter and the medium pressure ultraviolet sterilizing device via the ballast pump and is circulated through the internal circulation pipe. When the UV sterilizer reaches a stable state, the DV4 valve is closed and the DV5 valve is opened to enter the normal ballast operation program. In the normal ballast operation program, the UV dose and the set dose are compared by calculating the seawater flow and UV intensity. The control unit will automatically adjust the UV intensity and the opening of the DV5 valve; at the same time, if the backwash filter pressure difference is too high, the backwash filter will automatically enter the self-cleaning program, as shown in Fig. 7.31.

3. **Ballast stop mode working interface**
When the ballast mode is completed and the system needs to be stopped, simply click on the "Stop System" option on the display and the system will automatically shut down the various devices and valves enter the stopped state, as shown in Fig. 7.32.

4. **Ballast stop mode description**
When the system enters the stop program, the DV4 valve is first opened and the DV5 valve is closed. At this time, the system enters the backwash filter and the medium pressure UV sterilizer cleaning procedure. After the cleaning process is completed, the medium-pressure ultraviolet sterilizer stops working. After the cooling time of 30 s has elapsed, the ballast pump stops working. The DV1 valve, DV2 valve, and DV4 valve are closed and the ballast stops working (this process takes about 10 min), and the system returns to standby.
Note In the process of starting the system, if the stop option is clicked accidentally or by mistake, the system will still start up to the normal operation mode and then automatically enter the stop program.

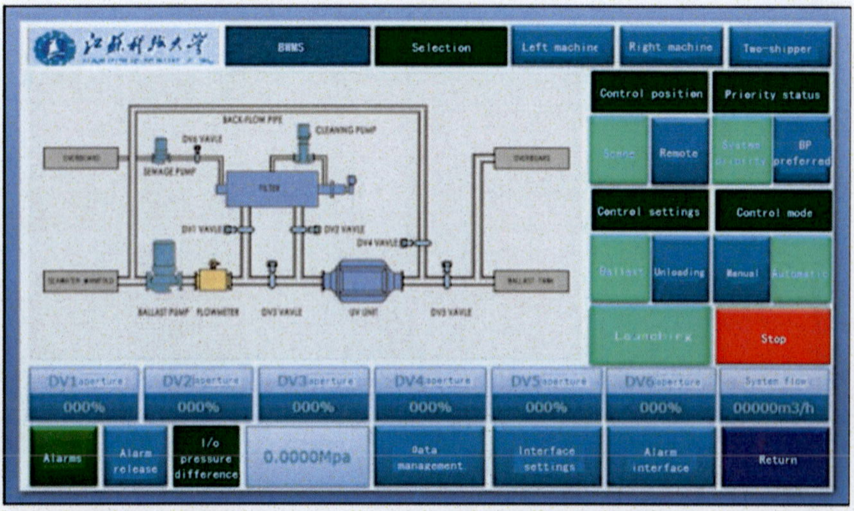

Fig. 7.32 Ballast stopping mode

5. **Unloading mode interface**
   After the system enters the unloading mode, its working interface is the same as the working interface of the ballast mode. However, the flow of the ballast water is different. The specific operation steps are as follows.

   (1) Turn the knob on the control panel to "Automatic" mode. Click on the "enter the system main interface" button on the display to enter the main interface of the system, as shown in Fig. 7.33.

   (2) Select "left machine", "right machine" or "dual machine" mode according to equipment and ship conditions. Select "Live" or "Remote" under the "Control Location" option. The selected item changes from dark green to light green, as shown in Fig. 7.34.

   (3) Select "System Priority" under the "Priority Status" option. The selected item changes from dark green to light green, as shown in Fig. 7.35.After the system enters the unloading mode, its working interface is the same as the working interface of the ballast mode. However, the flow of the ballast water is different. The specific operation steps are as follows.

   (4) Select "Uninstall" under the "Control Settings" option. The selected item changes from dark green to light green, as shown in Fig. 7.36.

   (5) Select "Automatic" under the "Control mode" option. The selected item changes from dark green to light green, as shown in Fig. 7.37.

   (6) At this time, the preparation of the system uninstall mode control interface is ready. Check the status of each unit and device in the system again and

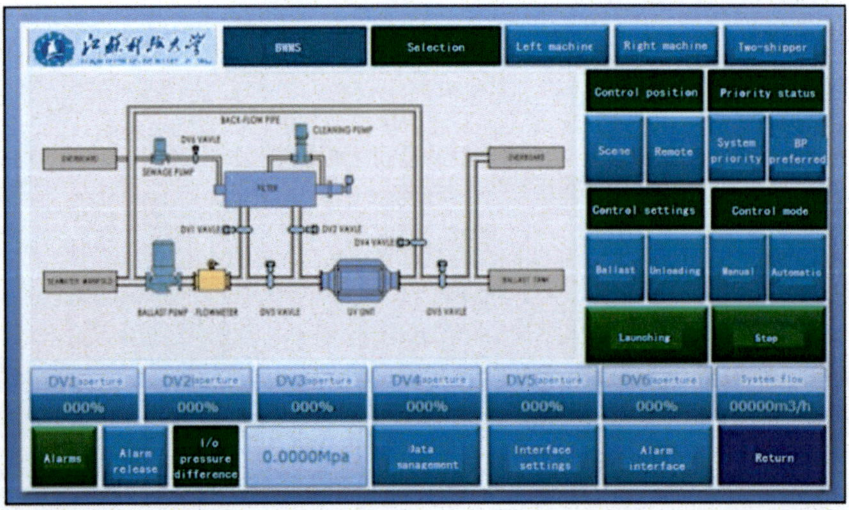

Fig. 7.33 Unloading management interface

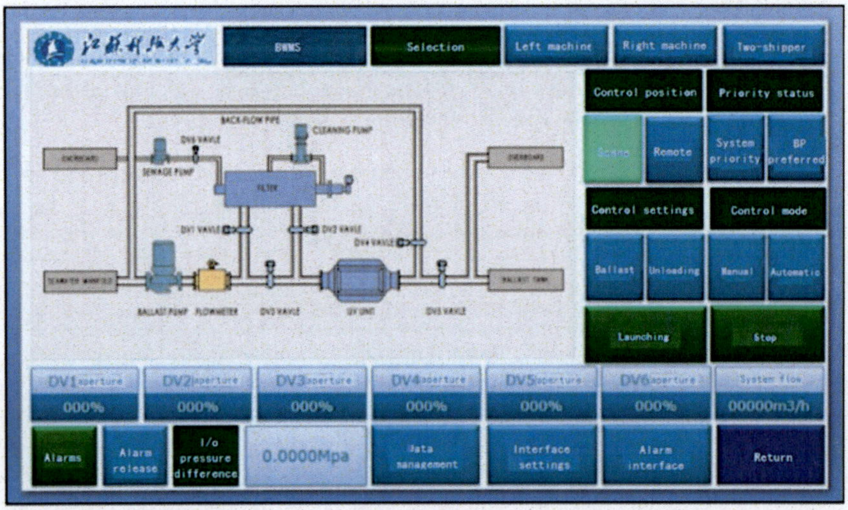

Fig. 7.34 Unloading operation mode selection

ensure that it is normal. Click on the system startup option and the system starts to run in uninstall mode. The selected item changes from dark green to light green, as shown in Fig. 7.38.

6. **Unloading mode description**
When the After the system is started, the DV3 valve and DV4 valve are opened. After the valvesareopened, the ballast pump starts. The seawater flows into the

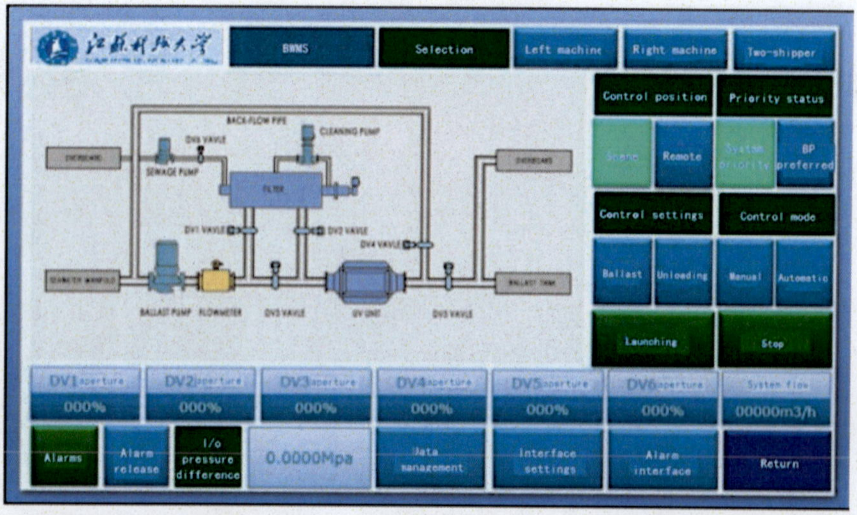

Fig. 7.35 Unloading priority mode

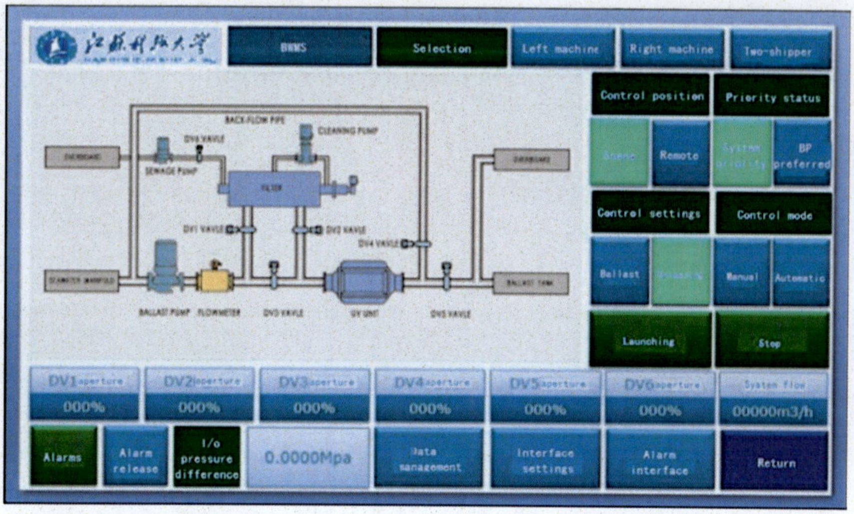

Fig. 7.36 Unloading control mode setting

medium-pressure ultraviolet sterilizing device via the ballast pump and is circulated through the return pipe. When the medium-pressure ultraviolet sterilizer reaches a stable state, the DV4 valve is closed and the DV5 valve is opened, and the normal unloading operation program is entered. When the normal operating procedure is unloaded, the resulting UV dose is compared with the set dose by

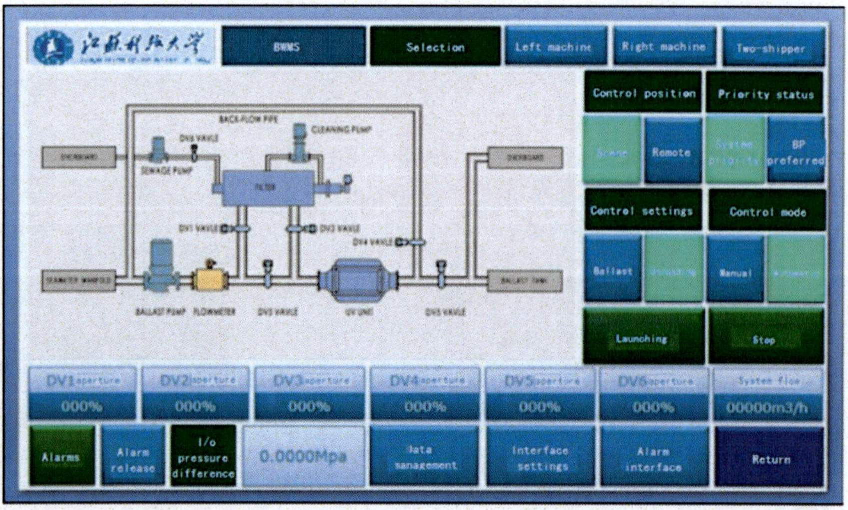

Fig. 7.37 Unloading automatic control mode

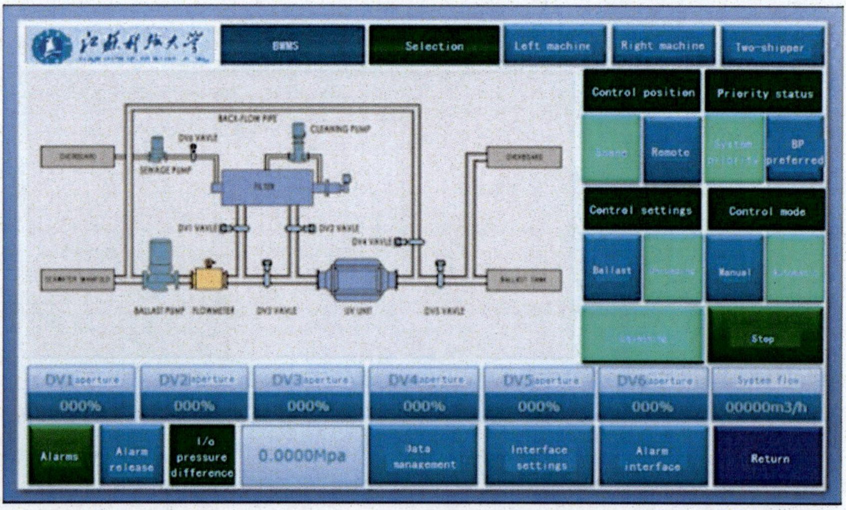

Fig. 7.38 System unloading mode

calculating the seawater flow and UV intensity. The control unit will automatically adjust the UV intensity and the opening of the DV5 valve, as shown in Fig. 7.39.

7. **Unloading stop mode**
unloading mode is completed and the system needs to be stopped, simply click on the "Stop System" option on the display. The system will automatically shut

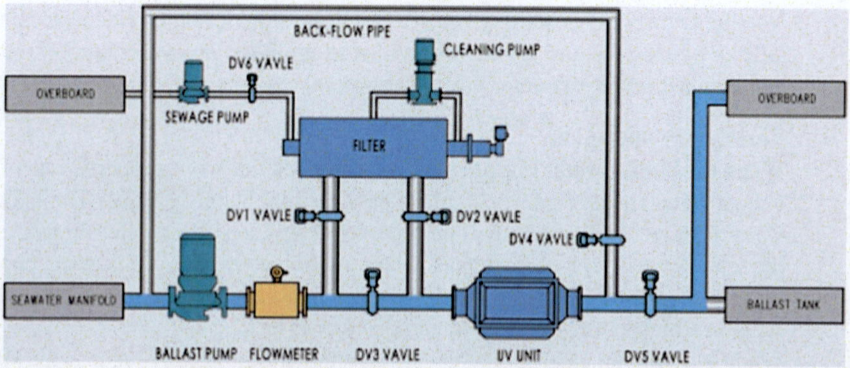

Fig. 7.39 Unloading mode flow diagram

down the various devices and valves and enter the stopped state, as shown in Fig. 7.40.

8. **Unloading stop mode description**
When the system enters the stop program, first the DV4 valve is opened and the DV5 valve is closed. At this point, the medium-pressure ultraviolet sterilizer stops working. After the 30-s cooling time is over, the ballast pump stops working. The DV3 valve and DV4 valve are closed and the ballast is stopped (this process takes about 10 min), and the system returns to standby.

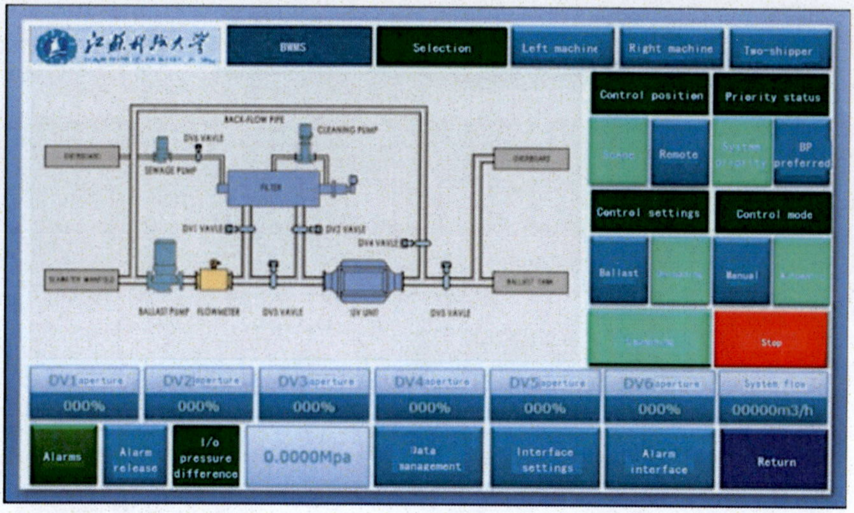

Fig. 7.40 Unloading stopping mode

Note In the process of starting the system, if the stop option is clicked accidentally or by mistake, the system will still start up to the normal operation mode and then automatically enter the stop program.

9. **Emergency operation**
   When the ballast water management system does not work properly, open the ballast water management system by-pass manual butterfly valve. The ballast water management system will be removed from the ballast water system and the seawater will bypass the ballast water treatment device. At this point, check the cause of the failure of the ballast water treatment device and eliminate it. After troubleshooting, normal operation can begin.
   When entering an emergency state, the system automatically records the start and end times.
   The working conditions of the system have been set at the factory. Testing staffcan modify and set the parameters according to actual needs, as shown in Fig. 7.41.

10. **User data parameter setting**
   Enter the main interface of the system, click on the "Data Management" option in the main interface of the system. The user can set the corresponding parameters and information. After setting, click the "Save" option, as shown in Fig. 7.42.
   To end the system operation, press the red "Stop" button on the control panel, the system will cut off the power (this process is a manual operation), that is, the system ends the work.

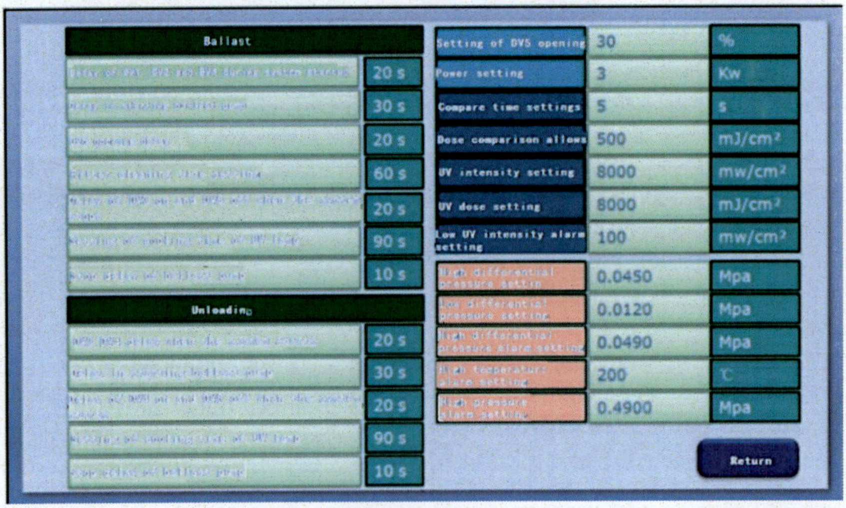

Fig. 7.41 Parameter modification and setting interface

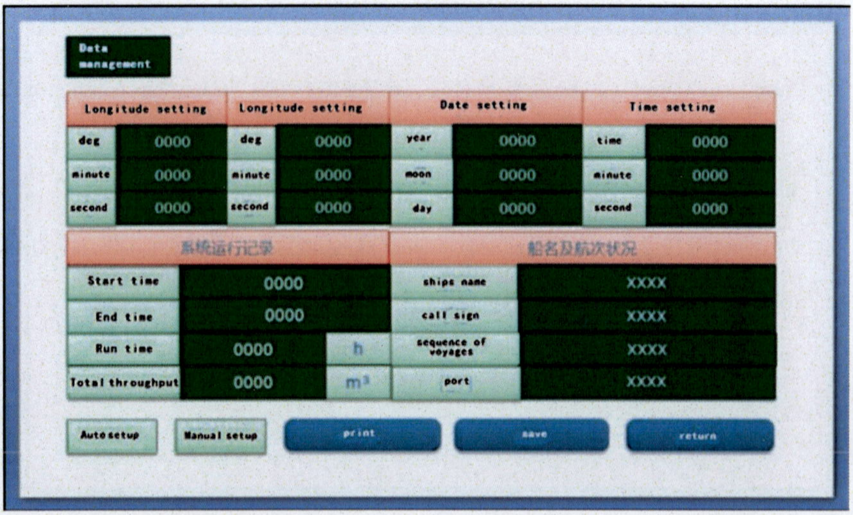

Fig. 7.42 Data management interface

When power is lost, the system stops operating. When the power is restored, the display returns to the main screen. The memory function of the system will automatically save the operating parameter data in the unpowered state. Just follow the normal ballast or unloading operation, as shown in Fig. 7.43.

Fig. 7.43 Ballast water management system after system recovery

References

1. Cheng-yan, Wang, Liu Xi-min and Ran Xiang-cheng et al. 2009. Preparation of nanosized TiO2 and its Applied in COD Detection. *China Environmental Testing* 25(3): 66–6. Beijing
2. Qingdao Shuangrui BalClor®. 2011. Ship Ballast Water Management System Operation and Maintenance Manual.Qingdao: Qingdao Shuangrui Marine Environment Engineering Co., Ltd.
3. JFE BallastAce®. 2013. Ship Ballast Water Management System Operation and Maintenance Manual. Japan: JFE Corporation.
4. YPBWMS®. 2013. Ballast Water Management System Operation and Maintenance Manual. Wenzhou: Zhejiang Yingpeng Marine Equipment Manufacturing Co., Ltd.
5. SEMENS S7–200 programmable controller [PC/OL]. 2000. Germany: Siemens.
6. KingView User Manual [PC/OL]. 2010. Beijing: Beijing Asia Control Technology Development Co., Ltd.

Chapter 8
Maintenance and Management of Ballast Water Management System

The ballast water management system, whether it is ultraviolet, inert gas or electrolytic or pharmacy, will encounter failures during the operation of the system and require the crew to carry out necessary maintenance. Due to the existence of power and high temperature, corrosion, noise, slippery and high and low and other dangerous conditionsaround the ballast water management system, therefore, the ballast water management system module units are affixed with the corresponding hazard warning labels. In addition, the maintenance of ballast water management systems developed according to different principles and methods varies from units to units and needs to be treated separately [1].

8.1 Ballast Water Management System Warning Labels

The warning labels surrounding the module units of ballast water management system are shown in Table 8.1.

8.2 Maintenance of Ballast Water Pretreatment Filtration Equipment

8.2.1 *Maintenance and Management of Backwash Filters*

1. **Backwash filter overview**

   The ballast water management system backwash filter is a fully automatic backwash filter made of SUS316L stainless steel. The seawater is filtered by coarse and fine filters. The reducer at one end of the backwash filter drives the drive shaft and the cleaning pipe to perform synchronous circular motion. The suction

N. Chen et al., *Development and Implementation of Ship BWMS*,
https://doi.org/10.1007/978-981-33-6865-1_8

Table 8.1 Warning labels around the BWMS module unit

Warning sign	Warning sign description
	Caution warning signs
	Charged Dangerous!
	All power must be cut off and marked before the equipment is operated, maintained, cleaned and overhauled. The power supply includes power distribution, or backup power
	Please wear UV protection mask when maintaining the medium-pressure ultraviolet disinfection device!
	Without protection, long-term ultraviolet radiation can burn eyes and skin. Unless you wear a UV protection mask provided in the operator's tool cabinet, direct viewing light tubes should be avoided
	Please wear protective gloves! Wear protective gloves when performing maintenance and inspections on the equipment
	Please wear earmuffs! Except for noise generated by the inert gas system air compressor, other ballast water management systems do not generate noise, but the operator should take care to protect the hearing from other possible noise in the ship
	Please wear safety boots! Wear safety boots when performing maintenance and overhaul on the system
	Be careful of corrosion! Cleaning fluids may cause uncomfort to the operator and avoid inhalation, ingestion or splashing on the eyes and skin. Wear overalls and rubber gloves
	Please wear a safety helmet! Please wear an appropriate helmet according to the safety regulations of the system equipment space
	Non-slip! System bases and other surfaces may be slippery when there is oil or water
	Prevent falling into the sky! Care should be taken when working near the system. Care should be taken to tighten the coupling bolts after removing or returning equipment or parts
	Be careful of slip! Be careful where you may be slipping before working on the system
3 ~	Three-phase AC!
~	Single phase AC!
I	On (Power)
O	Off (power)
	Ground protection!
	Please refer to the electrician's manual for maintenance!

(continued)

Table 8.1 (continued)

Warning sign	Warning sign description
	Be carefull of hot! Wait until the electrical equipment and the lamp tube cool down before touching. Danger of burns. Stainless steel surface or exposed to sunlight. Do not touch the stainless steel surface of the equipment unless wearing gloves

pipe onthe cleaning pipe andthe cleaning pipe itself are installed with a certain number of suction nozzles and cleaning nozzles, respectively. The differential pressure transmitter is connected to the inlet and outlet of the backwash filter. It is controlled by the system control center and checks the pressure difference between the inlet and outlet of the backwash filter.

The backwash filter pre-filters seawater throughout the system. With the double filtration of coarse and fine filters, solid impurities larger than 50 microns in seawater are excluded, making the seawater with good clarity and transparency, as shown in Fig. 8.1.

The performance indicators of the backwash filter are shown in Table 8.2.

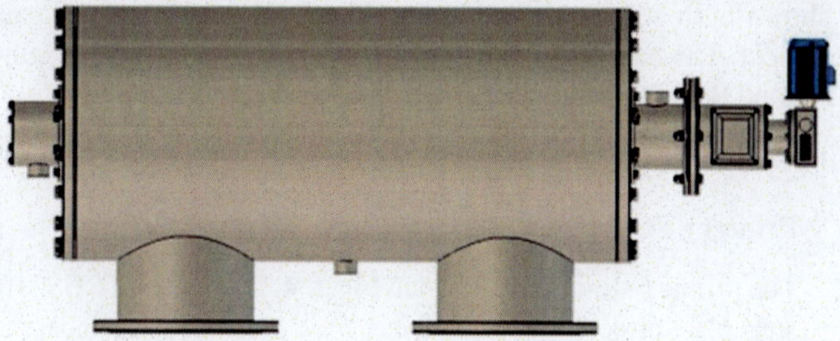

Fig. 8.1 Backwash filter

Table 8.2 Performance index of backwashing filter

Performance indicator	Performance index parameters
Minimum working pressure	2 bar (30 psi)
Maximum working pressure	10 bar (150 psi)
Filter pressure loss	0.1 bar (1.5 psi)
Maximum working water temperature	65 °C (149 °F)
Filtration accuracy range	20–50 μm
Water required for self-cleaning	Depends on the device model
Self-cleaning time	30 s
Self-cleaning control method	Differential pressure, time and manual

Table 8.3 YP≋BWMS® BWMS backwashing filter type

Model	Import and export D_1	Design flowm^3/h	D mm	L mm	L_1 mm	L_2 mm	H mm
YP-JUST-200F	DN250	300	560	755	1296	1989	714
YP-JUST-500F	DN300	500	660	855	1521	2214	814
YP-JUST-800F	DN350	800	750	1000	1250	2563	906

The installation method includes the online type and the in-line type. The backflush filter can be installed horizontally according to the working conditions on site.

Taking the YP≋BWMS® ballast water management system developed by Jiangsu University of Science and Technology as an example, the model parameters of backflushing filter for ship ballast water management system are shown in Table 8.3.

The dimensions of the YP≋BWMS® ballast water management system backwash filter are shown in Fig. 8.2.

YP≋BWMS® Ballast Water Management System Backwash Filter Pressure Drop is shown in Fig. 8.3.

As shown in the figure above, all baselines are under normal operation of the equipment. The data is collected for the water in free air with a temperature of 20 °C, and the kinematic viscosity $\gamma = 1\ mm^2/s$.

2. **Main components in the backwash filter**

(1) **Primary Strainer**

The primary strainer is shown in Fig. 8.4. It is made of SUS316L stainless steel plate directly punched. It has good pressure resistance, surface filtration, can be used repeatedly, and has a long service life. The parameters are shown in Table 8.4.

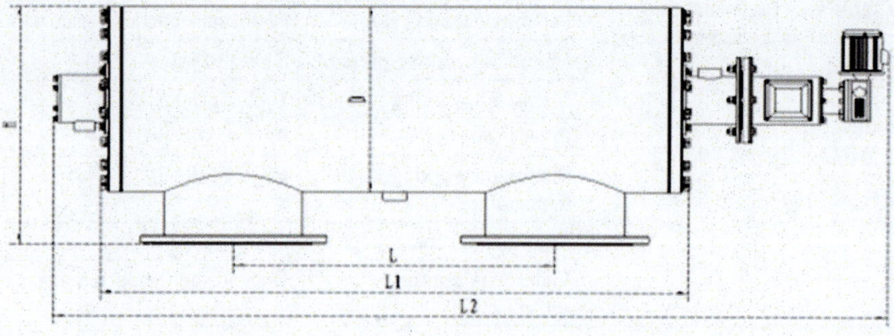

Fig. 8.2 Backwash filter size

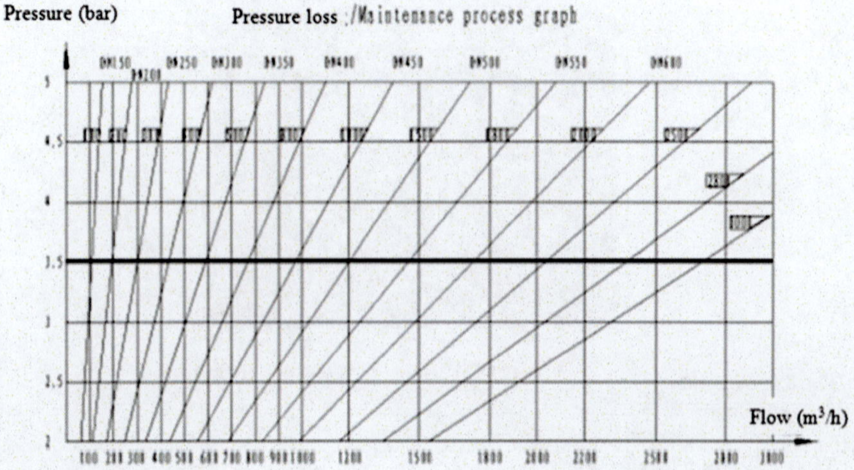

Fig. 8.3 Head filter loss

Fig. 8.4 Primary strainer

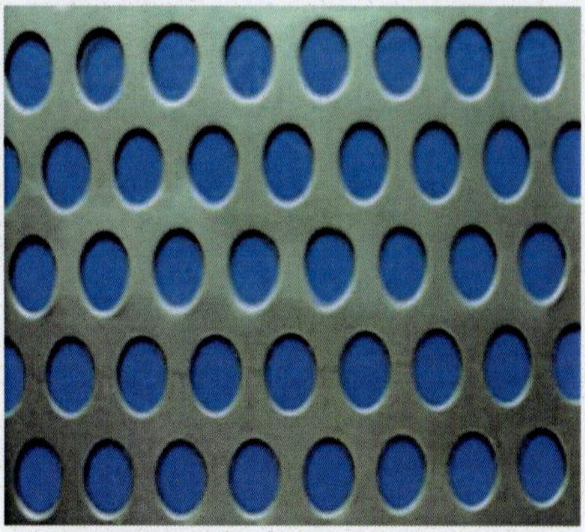

Table 8.4 Primary strainer parameters

Filtration accuracy	2 mm
Materials	SUS316L stainless steel
Filter length	Depends on backwash filter model
Filter diameter	Depends on backwash filter model
Filter direction	From outside to inside

(2) **Fine filter**.

The fine filter is shown in Fig. 8.5, a type ofmulti-layer sintered metal mesh, made ofmulti-layer metal mesh. It is manufactured by a special

Fig. 8.5 Fine filter

Table 8.5 Fine filter parameters

Filtration accuracy	40 μm
Materials	SUS316L stainless steel multilayer sintered mesh
Filter length	Depends on backwash filter model
Filter diameter	Depends on backwash filter model
Filter direction	From inside to outside

lamination press and vacuum sintering process.With higher mechanical strength and overall rigid structure, the parameters are shown in Table 8.5.

(3) **Proximity sensor**.

The proximity sensor is mounted 1 mm from the trip disc. Its appearance and technical parameters, and wiring diagrams are described in detail in Sect. 7.1 "Test Sensors for Ballast Water Management Systems", and will not be repeated here.

(4) **Reducer**

The reducer is composed of a motor and a reduction gearbox. Its role is when the filter is cleaned, the motor drives the worm gear and the reduction gear set in the reduction gear box to decelerate, and the drive screw drives the sewage suction tube to rotate,so that the suction head rotates along the left and right sides of the filter wall and moves to complete the filter cleaning. The appearance of the reducer is shown in Fig. 8.6. The parameters are shown in Table 8.6.

(5) **High Pressure Cleaning Pumps Overview**

The high-pressure cleaning pump in the ballast water management system YP≋BWMS®is a light vertical multi-stage centrifugal pump connected to the clean water chamber of the backwash filter. When the backwash

Fig. 8.6 Reducer

Table 8.6 Reducer parameters

Box material	Aluminum alloy
Center distance	63 mm
Reduction ratio	100
Permitted radial loading force	Depends on the backwash filter model

filter is used to clean the filter, high pressure cleaning water is provided for the backwash filter.Its appearance is shown in Fig. 8.7, and its technical parameters are shown in Table 8.7.

Fig. 8.7 High pressure cleaning pump

Table 8.7 Parameters of high pressure pump

Liquid temperature	−15~ + 120 °C
Connection method	Flange connection
Motor power	Depends on backwash filter model
Motor speed	Depends onbackwash filter model
Limited Data	Depends onbackwash filter model
Materials	SUS316L stainless steel
Weight	Depends onbackwash filter model

(6) **Sewage pump**

The sewage pump in the ballast water management system YP≈BWMS® is a single-stage centrifugal pump connected to the sewage chamber of the backwash filter. When the backwash filter cleans the filter, the sewage in the backwash filter is discharged. Its appearance is shown in Fig. 8.8. The technical parameters are shown in Table 8.8.

(7) **Differential pressure transmitter**

The differential pressure transmitter is installed between the inlet and outlet of the backwash filter. Its shape and technical parameters are also described in detail in Chap. 7, Sect. 7.1 "Bearing Sensors for Ballast Water Management Systems" and will not be repeated here.

3 **The working principle of backwash filter**

The backwash filter in YP≈BWMS® has a filtration accuracy of 40 μm and can automatically clean the filter.The seawater enters the coarse filter through the

Fig. 8.8 Sewage pump

Table 8.8 Sewage pump main parameters

Motor speed	2900 r/min
Motor power	1.5 kW
NPSH	2.3–2.5 NPSH
Flow	8.8–6.3 m^3/h
Pump head	17.8–21.2 m
Effectiveness	49–57%
Weight	38 kg

backwash filter inlet via a ballast pump, and is initially filtered by a coarse filter to filter impurities with a diameter greater than 2 mm.The seawater flows from the coarse filter to the fine filter for secondary filtration, and the impurities larger than 40 μm in diameter are filtered out. The twice-filtered seawater is discharged from the outlet of the backwash filter to complete the seawater filtration process.When impurities accumulate on the fine filter to reach a certain amount, a pressure difference will form between the inlet and outlet of the backwash filter when the pressure difference reaches a predetermined value.The speed reducer, sewage pump, drain solenoid valve and high-pressure cleaning pump controlled by the PLC start at the right time. The speed reducer drives the cleaning pipe, the suction pipe, and the water pipe to perform a 360-degree circular motion along the fine filter in the backwash filter cavity, and sprays the high-pressure cleaning water in the clean water cavity from the nozzle to the fine filter net. The suction nozzle on one side sucks impurities, such as sediment etc., adhering to the fine filter screen into the suction pipe, and finally the impurities are discharged from the sewage chamber through the sewage pump.At this point, the backwash filter completes a self-cleaning process. It should be noted that seawater filtration and self-cleaning of the filter are performed at the same time, and the two do not interfere with each other.

4 Maintenance instructions for the backwash filter

Before and after the maintenance of the backwash filter in YP≈BWMS®, all installation components and working media must be checked to avoid unintended operation. When carrying out the maintenance and repairwork, please perform the following five steps:

(1) Switch off the main power switch;
(2) Place warning signs to avoid restarting;
(3) Check the input and output lines connecting the units and place the valve opening and closing signs on the valves;
(4) After the maintenance work is completed, check that all safety devices are operating properly;
(5) Check that all connections (threads and fittings) are tightly fastened.

The following precautions must be taken when installing the backwash filter:

(1) When installing, please confirm the inlet and outlet of the backwash filter (marked on the back flush filter housing);
(2) After installing the backwash filter in place, check that the control line and the components are tightly connected;
(3) Check whether the control pipeline is crushed or deformed. If so, please handle it. Otherwise, it will directly affect the normal operation of the backwash filter.
(4) Check the drain pipe from the drain valve:
   ① Check whether the installation of the sewage pipeline is horizontal and minimize the detour.
   ② Make sure that the outlet of the sewage pipeline is in direct contact with the atmosphere. If there is pressure in the pipeline, the drain valve will not close during the anti-cleaning process and will directly affect the normal operation of the backwash filter.
   ③ The pipeline from the discharge end of the solenoid valve is also in direct contact with the atmosphere and no pressure is allowed in the pipeline.
(5) Make sure that the circuit and cable connections are well-water proofed and that they are not exposed to water.
(6) After completing the above confirmation, if a valve is installed at the outlet of the backwash filter, open it, slowly open the inlet valve, confirm that there is no leakage, and then turn on the power.
(7) After the installation is complete or there is a problem during installation, contact a professional technician to debug or answer the problem.

The operating conditions of the backwash filter are shown in Table 8.9.
If the on-site working conditions do not match the above conditions, please contact the manufacturer or agent to finda solution.

5. **Debuging of backwash filter** [1, 2]

(1) Carefully check the connection of the backwash filter control line.
(2) Slowly open the inlet valve of the backwash filter (if the backwash filter is fitted with an outlet valve, open it first) and verify that there is no abnormal working pressure.

Table 8.9 Working conditions of backwashing filter

Working environment	Working parameters	Working environment	Working parameters
Ambient temperature	≥ 0 °C	Minimum working pressure	2 bar
Operating temperature	≤ 65 °C	Maximum working pressure	10 bar

(3) Check for leaks at each connection of the back flush filter.
(4) Check the backwash filter accessory equipment (sewage pump, high pressure cleaning pump, etc.) to ensure that it works properly.

The backwash filter in YP≋BWMS® consists of the following seven units, as shown in Fig. 8.9. The names of the components are shown in Table 8.10.

(1) **Sewage chamber unit**
The sewage chamber is an important part of the filtration system. It is the sewage collecting area of the backwash filter. The sewage outlet connects the pipelines and discharges the sewage outboard to ensure that the backwash filter operates in a good condition. Its breakdown structure is shown in Fig. 8.10, and the names of its components are shown in Table 8.11.

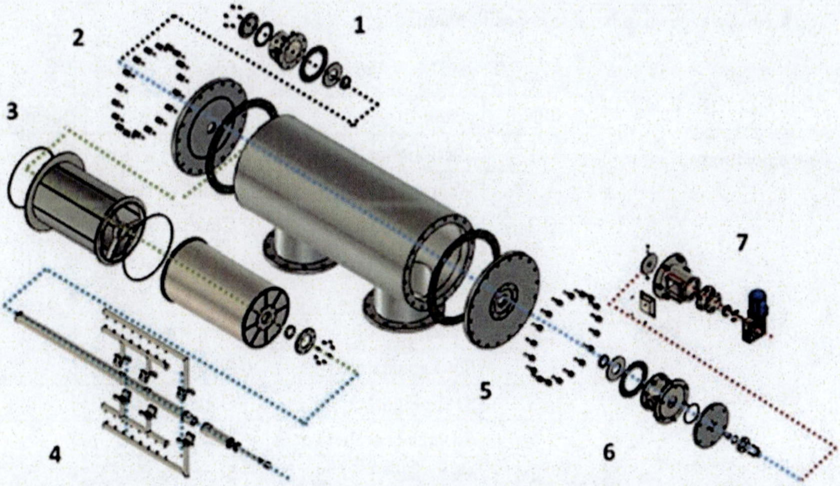

Fig. 8.9 Backwashing filter decomposition structure in YP≋BWMS®

Table 8.10 Anti-filter components details in YP≋BWMS®

No	Name	Quantity
1	Sewage chamber unit	1
2	Backwash filter inlet cap	1
3	Backwash filter screen assembly	1
4	Suction pipe, cleaning pipe assembly	1
5	Back flush filter outlet cover	1
6	Clear water chamber	1
7	Reduction mechanism assembly	1

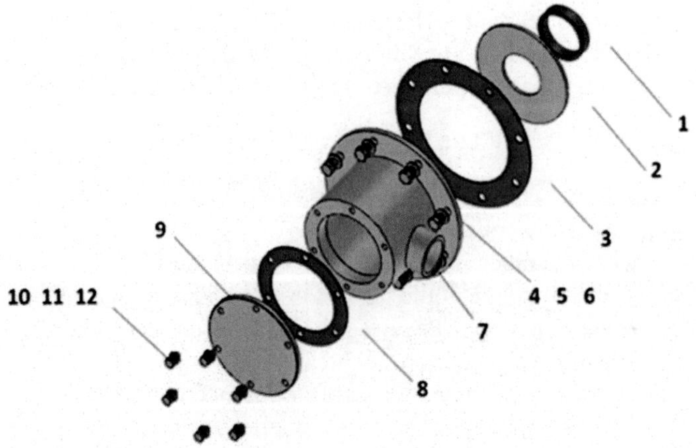

Fig. 8.10 Sewage chamber decomposition structure

Table 8.11 The details of sewage chamber decomposition structure

No	Name	Materials	Quantity
1	Suction pipe wear ring	Teflon	1
2	Wear ring gland	Stainless steel	1
2	Sealing gasket	Cork rubber	1
4	Flat gasket	Stainless steel	8
5	Spring washer	Stainless steel	8
6	Fastening bolt	Stainless steel	8
7	Sewage chamber	Stainless steel	8
8	Sealing gasket	Cork rubber	1
9	Sewage chamber cover	Stainless steel	1
10	Flat gasket	Stainless steel	6
11	Spring washer	Stainless steel	6
12	Fastening bolt	Stainless steel	6

According to the "decomposition diagram of the sewage chamber unit", component #7 in the diagram is a sewage chamber. When disassembling and installing the sewage chamber unit, please note the following:

① Close the backwash filter inlet and outlet valves and disconnect the power supply.
② Make sure the back flush filter drains water before servicing.
③ Remove the cavity cover plate bolts and washers 10,11, and 12.
④ Remove the sewage chamber cover and spacers 9 and 8.
⑤ Check and clean deposits in the sewage chamber.
⑥ Refill sewage tank cover plate and gasket.

⑦ Open the inlet and outlet valves of the backwash filter and turn on the power.
⑧ Check for leaks.
⑨ During disassembly, remember the number of parts removed and visually inspect them for proper preservation.

When the suction pipe wear ring is replaced, it should be based on the "decomposition diagram of the sewage chamber unit". Component #1 is the wear ring of the suction pipe.

① Close the backwash filter inlet and outlet valves and disconnect the power supply.
② Make sure the backflush filter drains water before servicing.
③ Remove the bolts and washers 4, 5, and 6 fixed to the inlet cover of the backwash filter.
④ Disassemble the sewage chamber 7.
⑤ Dismantle the sewage chamber gasket 3.
⑥ Take out wear ring gland 2.
⑦ Remove the wear ring and replace it with a new wear ring.
⑧ Put on wear ring gland 2.
⑨ Tighten the fastening bolts in the fixing the sewage chamber and press the wear ring tightly to 4, 5, and 6.
⑩ Open the inlet and outlet valves of the backwash filter and turn on the power.
⑪ Check for leaks.
⑫ During disassembly, remember the number of parts removed and visually inspect them for proper preservation.

(2) **Backwash filter inlet end cover**

The maintenance and installation of the inlet end cover of the backwash filter is shown in Fig. 8.11. The names of the components are shown in Table 8.12. When the inlet end cover of the backwash filter is disassembled and installed according to the "retroreflection filter inlet end cover exploded view", the following matters should be noted:

① Close the backwash filter inlet and outlet valves and disconnect the power supply.
② Make sure the backwash filter drains water before servicing.
③ Remove the backwash filter inlet cover and washers 1, 2, 3, 4, and 5 in sequence.
④ Check the use of seals and replace if necessary.
⑤ Install the bolts and washers of the back cover of the backwash filter.
⑥ Install the backwash filter inlet end cover 2.
⑦ Open the inlet and outlet valves of the backwash filter and turn on the power.
⑧ Check for leaks.

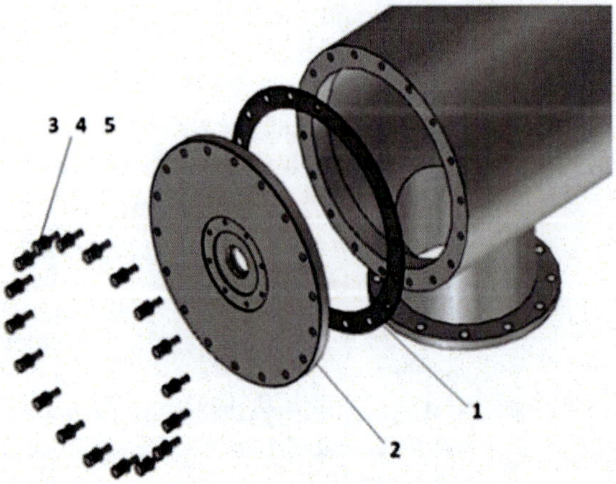

Fig. 8.11 Disassembly diagram of filter inlet end cover

Table 8.12 Filter inlet end cover parts details

No	Name	Materials	Quantity
1	Backwash filter cover gasket	Cork rubber	1
2	Backwash filter inlet end cover	Stainless steel	1
3	Flat gasket	Stainless steel	18
4	Spring washer	Stainless steel	18
5	Fastening bolt	Stainless steel	18

⑨ During disassembly, visually inspect the removed parts and keep them properly.

(3) **Backwash filter screen assembly**

The backwash filter screen assembly consists of a primary strainer and a fine filter, as well as a screen support ring and a fine filter seal kit, as shown in Fig. 8.12. The various components are shown in Table 8.13.

① **primary strainer**

The primary strainer is made of SUS316L stainless steel orifice plate and reinforced support welded with a filtration accuracy of 2 mm. The primary strainer is at the inlet of the backwash filter, intercepts the larger impurities in the water source, and completes the initial filtration of the seawater so that the fine filter can perform better filtration.

The primary strainer is disassembled and installed, according to the "retro-filter filter screen assembly exploded view", and the component #3 is the primary strainer. During installation and maintenance, the following matters shoud be noted:

I Close the inlet and outlet valves of the backwash filter and disconnect the power supply.

II Ensure that the backwash filter drains water before servicing.

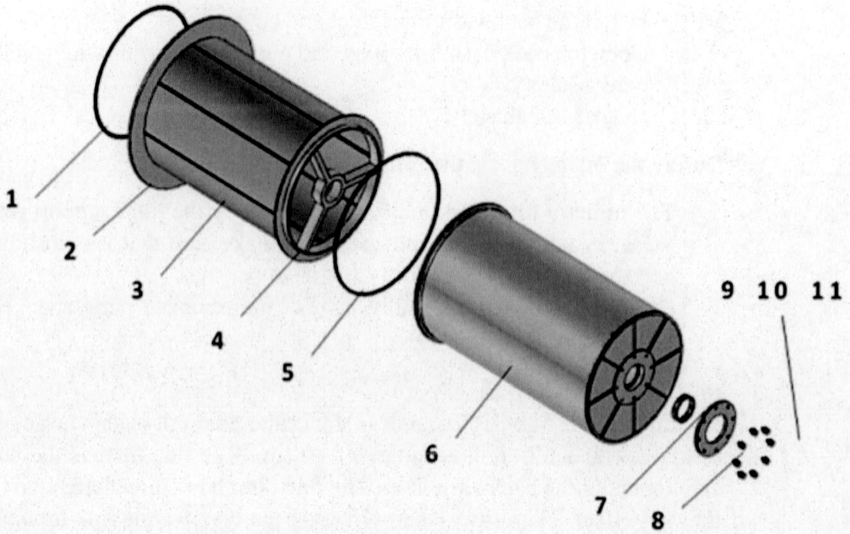

Fig. 8.12 Disassemble filter screen assembly diagram

Table 8.13 Filter screen component details

No	Name	Materials	Quantity
1	primary strainer sealing ring	Cork rubber	1
2	primary strainer positioning ring	Stainless steel	1
3	primary strainer	Stainless steel	1
4	Filter support ring	Stainless steel	1
5	"O" ring	Fluororubber	1
6	Fine filter	Stainless steel	1
7	Sealing ring	Teflon	3
8	Fine filter gland	Stainless steel	1
9	Flat gasket	Stainless steel	8
10	Spring washer	Stainless steel	8
11	Fastening bolt	Stainless steel	8

III Remove the backwash filter inlet end cover.

IV Remove the primary strainer seal 1 from the back groove in the back cover of the backwash filter.

V Remove the primary strainer from the backwash filter chamber 3.

VI Place the overhauled primary strainer or new one into the strainer assembly.

VII Check to see if the primary strainer seal is embedded in the groove on the back cover of the backwash filter.

VIII Install the backwash filter inlet end cover.
IX Open the backwash filter inlet and outlet valves and switch on the power supply.
X check for leakage.

During the maintenance, the following matters should be noted:

I The primary filter and fine filter are fixed by the filter support ring. When installing the primary strainer, make sure that it is installed horizontally in the backflush filter chamber.
II During disassembly, visually inspect the removed parts and keep them properly.

② **Fine filter**

The fine filter is SUS316L stainless steel sintered mesh made viavacuum welding with a filtration accuracy of 40 μm. The fine filter is the core component of the backwash filter. The fine filter has a direct effect on the filtering effect. Therefore, when selecting the type for the first time, the filter must be selected according to the requirements.

As for the fine filter disassembly and installation, please proceed according to "backwash filter screen assembly exploded view" shown, component #6 is a fine filter. During maintenance and installation, the following matters should be noted:

I Close the inlet and outlet valves of the backwash filter and disconnect the power supply.
II Ensure that the backwash filter drains water before servicing.
III Open the drive cavity cover and remove the trip disc attachment bolts (where the drive is connected), so that the drive shaft is separated from the drive shaft. See "Removal and Installation of Speed Reducer Components".
IV Remove the backwash filter outlet cover (water chamber end).
V Carefully remove the "suction tube and wash tube assembly".
VI Remove the primary strainer according to the section "Disassembly and installation of the primary strainer" in this section.
VII Remove the filter support ring and the fine filter "O" ring 4 and 5.
VIII Remove the "Cleaning Tube Assembly" and remove the "Cleaning Tube Assembly" from the backflush filter outlet.
IX Remove the long suction pipe from the inlet side of the backwash filter (rotate the suction pipe manually until it is pulled out smoothly from the chamber end).
X Remove the fine filter 6 from the inlet side of the backwash filter.
XI Check fine filter "O" rings and seals, replace 5, 7 if necessary.
XII Reassemblethe fine filter and fine filter "O" rings in the slots of the backwash filter housing.

XIII Refer to the installation of the primary strainer and the removal of the primary strainer 2.

XIV Ensure that the primary strainer seal is embedded in the groove on the back cover of the backwash filter, and the back cover of the backwash filter is replaced.

XV The Outlet is equipped with "suction pipe and cleaning pipe assembly".

XVI The Havea backflush filter outlet end cover.

XVII Eeassemble the trip disc to ensure that the transmission short axis and the transmission long axis are reliably connected. See "Removal and Installation of Speed Reducer Assembly".

XVIII Refit drive chamber end cover.

XIX Open the inlet and outlet valves of the backwash filter and turn on the power.

XX Check for leaks.

In addition, during installation and maintenance, the following matters should be noted:

I Check the "O" ring and packing at the end of the fine filter and replace it if necessary.

II When assembling the fine filter, the fine filter should be installed horizontally to avoid colliding with the suction pipe and the cleaning pipe assembly.

III After installing the filter assembly, the suction pipe and the cleaning pipe assembly can be freely rotated in the backwash filter chamber, and the filter screen does not touch the nozzle and the suction nozzle, and there is no abnormal noise. (The nozzle is about 2–3 mm from the outer surface of the fine filter, and the suction nozzle is about 2–2.5 mm from the inner surface of the fine filter.)

(4) **Suction pipe and cleaning pipe assembly**

The backflushing filter suction pipe and the cleaning pipe assembly are mainly composed of a sewage suction pipe and a cleaning pipe, and a support main pipe kit for the sewage suction pipe and the cleaning pipe. The end of the main pipe connected with the sewage suction pipe is connected with a sewage pump and connected with the cleaning pipe. One end of the pump is connected to the high-pressure cleaning pump, and the two sides are not conductive, as shown in Fig. 8.13. The components are shown in Table 8.14.

① **Cleaning pipe assembly**

The cleaning pipe assembly is made of stainless steel. Its role is to provide cleaning water to the fine filter in the backwash filter. During the cleaning process, high pressure cleaning water is sprayed on the outer surface of the fine filter. During this process, the reduction gear drives it to perform axial and circular motion.

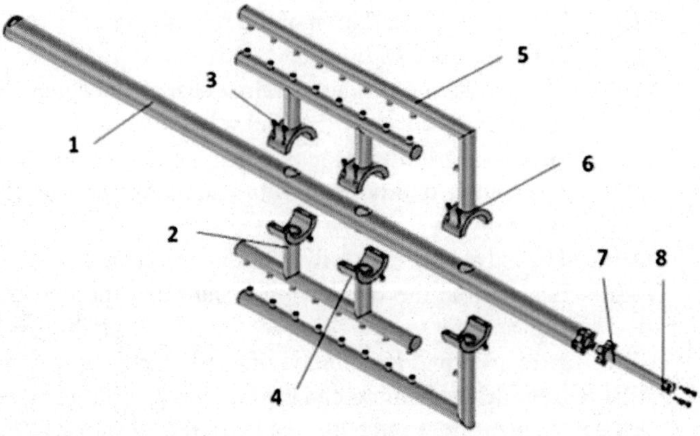

Fig. 8.13 Decomposition diagram of sewage suction and cleaning pipe assembly

Table 8.14 Details of sewage suction and cleaning pipe assembly

No	Name	Materials	Quantity
1	Suction pipe	Stainless steel	1
2	Suction branch	Stainless steel	2
3	Fastening bolt	Stainless steel	12
4	Suction pipe U-shaped split hole connector	Stainless steel	4
5	Cleaning branch	Stainless steel	2
6	Cleaning branch U-shaped split hole connector	Stainless steel	2
7	Transmission shaft	Stainless steel	1
8	Fastening bolt	Stainless steel	4

The cleaning pipe assemblies for different types of backwash filter are different and its length depends on the model.

Cleaning pipe assembly removal, installation, and maintenance, should be carried out according to the "absorbed pipe and cleaning pipe assembly exploded view" shown. Component #5 is the cleaning pipe assembly.During maintenance, the following matters should be noted:

I Close the inlet and outlet valves of the backwash filter and disconnect the power supply.

II Ensure that the backwash filter drains water before servicing.

III Open the drive chamber cover and remove the trip disc coupling pin (drive connected here), so that the drive shaft and the drive shaft are separated. Please refer to"Removal and Installation of Speed Reducer Components".

IV Remove the backwash filter outlet end cover (clean water chamber end).

V Remove the cleaning pipe U-shaped split hole connector fastening bolts and remove component #7 from the transmission stub.
VI Slowly remove the purge manifold assembly from the outlet end of the backwash filter and take out component 5.
VII Check the use of various parts and seals. Replace them if necessary.
VIII Reassembly cleaning branch assembly 5.
IX Fasten the purge pipe U-type split hole connector 6 from the outlet end of the backwash filter.
X Check the centering and fastening conditions of the long suction pipe, the cleaning branch pipe and the rotating shaft 1, 5, and 7.
XI Install back flush filter outlet cover (clean water chamber end).
XII reassembled the travel disc to ensure that the transmission short axis and the transmission long axis are reliably connected. Please refer to"Removal and Installation of Speed Reducer Assembly".
XIII Reassemblethe transmission chamber cover.
XIV Open the inlet and outlet valves of the backwash filter and turn on the power.
XV Check for leakage.

In addition, during the installation and maintenance, the following matters should be noted:

I The cleaning pipe assembly requires good concentricity during installation. Therefore, the collision of the cleaning pipe assembly must be avoided during transportation and installation.
II After the cleaning pipe assembly is reinstalled, make sure that it is free to rotate in the backwash filter chamber without any jamming.

② **Suction pipe assembly**
The material of the suction pipe assembly is stainless steel, and its function is to complete the suction of impurities on the surface of the fine filter screen during the anti-washing process of the backwash filter. In the process, the reducer drives it to perform axial and circular motions.
Different types of backflushing filters have different suction pipe assemblies, and their length depends on the model.
For sewage suction pipe assembly removal and installation, please proceed according to "absorbed pipe and cleaning pipe assembly exploded view" shown. Component #2 is the sewage suction pipe assembly, During installation and maintenance, the following matters should be noted:

I Close the inlet and outlet valves of the backwash filter and disconnect the power supply.
II Ensure that the backwash filter drains water before servicing.
III Open the drive chamber cover and remove the trip disc coupling pin (drive connected here), so that the drive shaft and the drive shaft are separated. Please refer to "Removal and Installation of Speed Reducer Components".

IV Remove the backwash filter outlet cover (water chamber end).
V Remove the backwash filter inlet end cover (sewage chamber end).
VI Remove the primary strainer from the inlet side of the backwash filter.
VII Remove the cleaning pipe U-shaped split hole connector fastening bolts, and remove component#7 from the transmission stub.
VIII Remove the cleaning branch (sucking pipe and cleaning pipe joints dismantling is completed) 5.
IX Slowly remove the long suction pipe from the inlet side of the backwash filter (rotate the suction pipe manually until it is pulled out smoothly from the chamber end).
X Check the usage of various parts and seals. Replace themif necessary.
XI Replacing the filter assembly, Please refer to "Filter assembly dismantling and installation".
XII Refill suction pipe assembly 2.
XIII Refill cleaning pipe assembly 5.
XIV fastens the wash tubing split hole connector 6 from the outlet end of the backwash filter.
XV Check the centering and fastening of the suction pipe assembly, cleaning branch pipe and rotating shaft, 1, 2, and 7.
XVI Install a backwash filter inlet end cover (sewage chamber end).
XVII Install a backwash filter outlet cover (clear water chamber end).
XVIII Reinstallthe travel disc to ensure reliable connection of the transmission short axis with the transmission long axis. Please refer to "Removal and Installation of Speed Reducer Assembly".
XIX Refitdrive chamber cover.
XX Open the inlet and outlet valves of the backwash filter and turn on the power.
XXI Check for leakage.

In addition, during installation and maintenance, the following matters should be noted:

I The suction pipe assembly requires good concentricity during installation. Therefore, during the transportation and installation process, the suction pipe assembly must be prevented from colliding.
II After the suction pipe assembly is reinstalled, ensure that the suction pipe assembly is free to rotate in the backwash filter chamber without any jamming.

(5) **Backflush filter outlet cover**

The outlet cover of the backwash filter is mainly used for sealing the end face of the filter, and at the same time, it provides support for the installation of the clean water chamber and the reducer, as shown in Fig. 8.14. Its components are shown in Table 8.15.

The removal and installation of the outlet cover of the backwash filter is performed according to the "Exploded View of the Backwash Filter Outlet". When it is necessary to inspect the backwash filter cleaning water pipe assembly, the outlet cover of the backwash filter is opened in turn according to the schematic diagram, and the cleaning water pipe assembly is overhauled.

① Close the backwash filter inlet and outlet valves and turn off the power supply.
② Make sure the backwash filter drains water before servicing.
③ Remove the nuts and washers 1, 2, 3, 4, and 5 of the back cover of the backwash filter in order.
④ Check the usage of seals and replace them if necessary.
⑤ Install the bolts and washers of the back cover of the backwash filter.
⑥ Install the backflush filter outlet cover 2.
⑦ Open the inlet and outlet valves of the backwash filter and turn on the power.
⑧ Check for any leakage.

In addition, during installation and maintenance, the following matters should be noted:

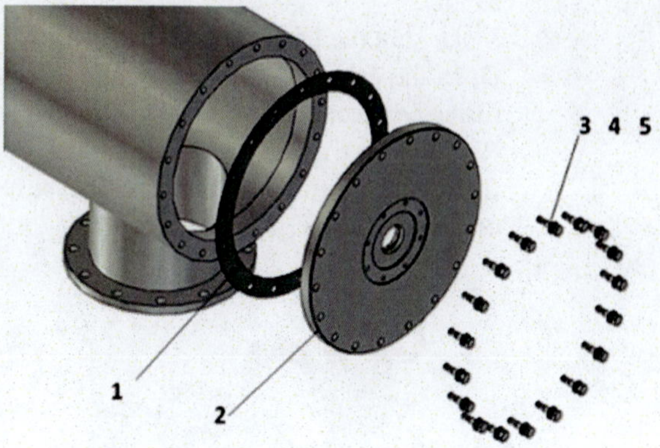

Fig. 8.14 Decomposition diagram of outlet cover of backwashing filter

Table 8.15 Details of outlet end cover parts of backwashing filter

No	name	Materials	Quantity
1	Backwash filter cover gasket	Cork rubber	1
2	Backwash filter outlet cover	Stainless steel	1
3	Flat gasket	Stainless steel	18
4	Spring washer	Stainless steel	18
5	Fastening bolt	Stainless steel	18

① The end cap of the cylinder can be removed during the inspection of the filter, keep shut down during maintenance, close the inlet and outlet valves of the backwash filter, and ensure that there is no water in the backwash filter.

② During disassembly, visually inspect the removed parts and keep them properly.

(6) **Clean water chamber**

The fresh water chamber unit of the backwashing filter is connected with the outlet end cover, and the short shaft of the speed reducer passes through the driving shaft to drive the cleaning main pipe to rotate, and the cleaning water input by the cleaning high pressure pump enters the cleaning pipe by the unit, and the high pressure flushing of the outer surface of the fine filter mesh is carried out,while supporting the reducer as shown in Fig. 8.15. The various components are shown in Table 8.16.

The removal and installation of the backwash water tank unit is performed as shown in Fig. 8.15. When it is necessary to repair the backwash water filter unit assembly or rotating the short shaft, press the "clear water chamber unit exploded view" diagram, and open the backwash filter water chamber unit in turn, to clean the water tank unit components. Component #4 is the water chamber.

① Close the backwash filter inlet and outlet valves and turn off the power supply.

② Make sure the backwash filter drains water before servicing.

③ Open the drive chamber cover and remove the trip disc attachment bolts (drive connection here).

④ Remove the trip disc coupling bolt to separate the transmission long axis from the transmission short axis. Please refer to "Removal and Installation of Speed Reducer Assembly".

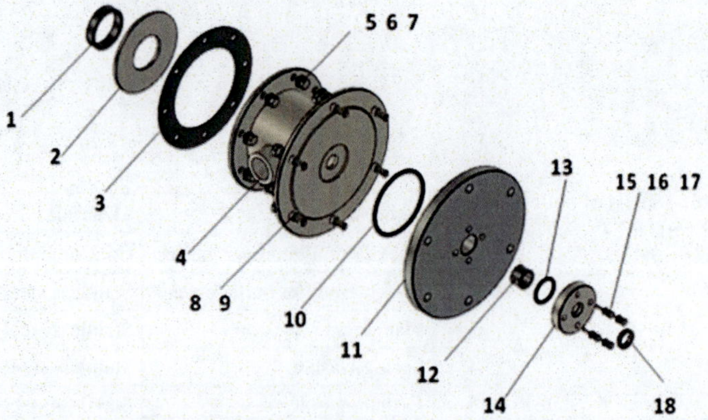

Fig. 8.15 Clean water unit decomposition diagram

Table 8.16 Details of clean water unit components

No	Name	Specification	Quantity
1	Cleaning pipe wear ring	Teflon	1
2	Wear ring gland	Stainless steel	1
3	Sealing gasket	Cork rubber	1
4	Water chamber	Stainless steel	1
5	Flat Washers	Stainless steel	8
6	Spring washer	Stainless steel	8
7	Fastening bolt	Stainless steel	8
8	Flat gasket	Stainless steel	6
9	Fastening bolt	Stainless steel	6
10	"O" ring	Fluororubber	1
11	Clean water chamber stuffing cap	Stainless steel	1
12	Packing	Teflon	3
13	"O" ring	Fluororubber	1
14	Clean water chamber gland	Stainless steel	1
15	Flat Washers	Stainless steel	4
16	Spring washer	Stainless steel	4
17	Fastening bolt	Stainless steel	4
18	Bidirectional piston seal	Polyurethane	1

⑤ Remove the connection flange between the speed reducing mechanism and the clean water chamber, and separate the speed reducing mechanism from the clean water chamber unit. Please refer to "Removal and Installation of Speed Reducer Components".
⑥ Check the usage of clean water chamber seals and two-way piston seals. Replace components 1, 2, 18 if necessary.
⑦ Replacement reducer and water chamber connection flange.
⑧ Reinstall the travel disc to ensure that the transmission short shaft and the transmission long shaft are reliably connected. Please refer to "Removal and Installation of Speed Reducer Components".
⑨ Restore the drive chamber cover.
⑩ Open the inlet and outlet valves of the backwash filter and turn on the power.
⑪ Check for any leakage.

In addition, during the installation and maintenance, the following matters should be noted:

① During maintenance, the system shall be shut down, the inlet and outlet valves of the backwash filter shall be closed, and no water shall accumulate in the backwash filter.

② During disassembly, remember that the removed parts are visually inspected and properly stored.

When the wear ring of the cleaning pipe is replaced, it needs to be performed according to the "decomposition diagram of the clean water chamber", and component #1 is the wear ring of the cleaning pipe.

① Close the backwash filter inlet and outlet valves and turn off the power supply.
② Make sure the backwash filter drains water before servicing.
③ Open the drive chamber cover and remove the trip disc attachment bolts (drive connection here).
④ Remove the trip disc coupling bolt to separate the transmission long axis from the transmission short axis. Please refer to "Removal and Installation of Speed Reducer Components".
⑤ Remove the connection flange between the speed reducing mechanism and the clean water chamber, and separate the speed reducing mechanism from the clean water chamber unit. Please refer to "Removal and Installation of Speed Reducer Components".
⑥ Remove the bolts and washers 5, 6 and 7 fixed to the outlet cover of the backwash filter.
⑦ Detach the water chamber 4.
⑧ Remove the water chamber seal 3.
⑨ Remove the clean water chamber wear ring gland 2.
⑩ Remove the wear ring and replace it with a new wear ring.
⑪ Put on wear ring gland 2.
⑫ Tighten the fastening bolts securing the fresh water chamber and press the wear rings 5, 6, 7.
⑬ Reassemble the reduction mechanism and the water chamber unit flange.
⑭ Reinstall the trip disc to ensure that the drive shaft and the drive shaft are reliably connected. Please refer to "Removal and Installation of Speed Reducer Components".
⑮ Reassemble the drive chamber cover.
⑯ Open the inlet and outlet valves of the backwash filter and turn on the power.
⑰ Check for any leakage.

In addition, during the installation and maintenance, the following matters should be noted:

① During maintenance, the system shall be shut down, the inlet and outlet valves of the backwash filter shall be closed, and no water shall accumulate in the backwash filter.
② During disassembly, visually inspect the removed parts and keep them properly.

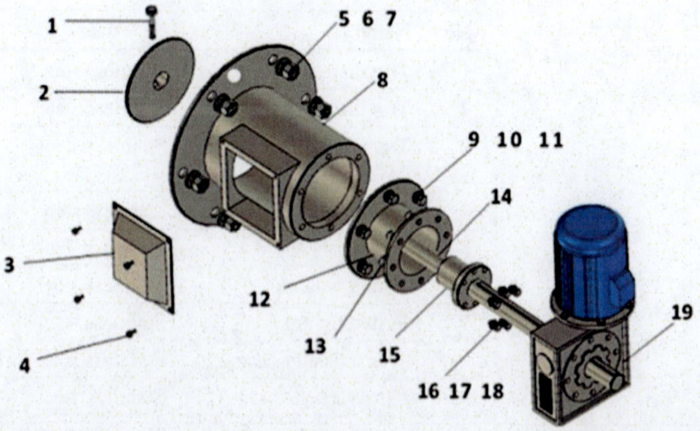

Fig. 8.16 Reducer component decomposition diagram

(7) **Reduction mechanism assembly**

The reduction mechanism is mainly used to reduce the speed of the motor, and the mechanism for rotating the backwash filter cleaning mechanism is driven by the short shaft. It is installed at the end of the clean water chamber, as shown in Fig. 8.16. The details of the components are shown in Table 8.17.

The removal and installation of the reduction mechanism components are performed according to the "exploded mechanism assembly reduction mechanism." When it is necessary to repair the fine filter, suction pipe, cleaning pipe, change the root of the clean water chamber, and replace the wear ring of the cleaning pipe, the drive shaft must be removed.

① Close the backwash filter inlet and outlet valves and turn off the power supply.
② Make sure the backwash filter drains water before servicing.
③ Remove the bolt 5 of the drive housing cover.
④ Remove the drive chamber cover (drive connected here) 4.
⑤ Remove the trip disc coupling bolts to separate the transmission long shaft from the transmission short shaft.
⑥ Check the usage of each part.
⑦ Replace the travel disc to ensure that the transmission short shaft and the transmission long shaft are reliably connected.
⑧ Repeat the drive chamber cover.
⑨ Open the inlet and outlet valves of the backwash filter and turn on the power.
⑩ Check for any leagage.

In addition, during the installation and maintenance, the following matters should be noted:

Table 8.17 Reducer components details

No	Name	Materials	Quantity
1	Trip disc coupling pin	Stainless steel	1
2	Trip disc	Stainless steel	1
3	Transmission chamber cover	Stainless steel	1
4	Fastening bolt	Stainless steel	4
5	Flat Washers	Stainless steel	6
6	Spring washer	Stainless steel	6
7	Fastening nut	Stainless steel	6
8	Transmission chamber	Stainless steel	1
9	Flat Washers	Stainless steel	6
10	Spring washer	Stainless steel	6
11	Fastening bolt	Stainless steel	6
12	Reducer connection base	Stainless steel	1
13	Fastening bolt	Stainless steel	8
14	Transmission shaft	Stainless steel	1
15	Screw nut	Stainless steel	1
16	Flat Washers	Stainless steel	6
17	Spring washer	Stainless steel	6
18	Fastening bolt	Stainless steel	6
19	Reducer	Aluminum alloy	1

① When installing a trip disc, make sure that the trip disc coupling pins are tight and prevent loosening.

② If not necessary, please do not remove the screw nut at all. Apply grease to the drive screw for good operation of the transmission mechanism.

③ Before the gearbox is put into operation, it is necessary to add the lubricating oil of appropriate viscosity, so that the friction between the gears should be reduced. In the case of high load and impact load, the reducer can fully exert its function.

④ When the reducer is first operated for about 200 h, the oil must be drained, rinsed, and then re-added with new lubricant to the center of the oil standard. If the oil level is too high or too low, the operating temperature may increase.

Lubricants used for gearboxes are shown in Table 8.18.

Table 8.18 Recommended lubricants grade

Lubricant type	Oil lubrication				Grease lubrication	Special lubricant	
Ambient temperature (°C)	0~+40	−15~+25	−30~+10	−50~20	−15~+40	−40~+80	−40~+10
ISO viscosity	VG220	VG100 VG150	VG68-46 VG32	VG22 VG15		VG220	VG32
BP	BP Energol GR-XP220	BP Energol GR-XP100	BP Energol GR-XP68	BP Energol LPT22	BP Energol HT-EP 00		
Mobil	Mobilgear 630	Mobilgear 629	Mobil D.T.E 15 M	Mobil D.T.E 11 M	Mobilplex 44	Mobil SHC 630	Mobil SHC 624
Shell	Shell Omala Oil 220	Shell Omala Oil 100	Shell Tellus Oil T32	Aero Shell Fluid 41	Shell Grease S3655		

8.2.2 *Maintenance and Management of Other Filtration Equipment*

The filter pre-treatment devices introduced in this book include ballast filters, ultrasonic cavity purifiers, ultrasonic cleaners, and membrane filters. Since these pre-treatment filtration devices are relatively simple in terms of the system structure of their devices in comparison tothe backwash filter, this book will only introduce the precautions of the several ballast water pre-treatment filtration devices during the maintenance process.

1 **The maintenance of swirl filter** [3]
Swirl filter is a device that uses the centrifugal force of ballast water entering the flow in the swirl chamber to separate impurities with a density higher than that of sea water. Its structure is relatively simple. The whole device is welded by SUS316L stainless steel, and the water flow enters from the vortex inlet. Sewage flows out from the underflow port, and clean water flows out of the upper overflow port without additional mechanical structure.
The inner wall surface of the chamber of the swirl filter is subject to cyclone separation and filtration, and is subject to long-term wear of the sand on the wall surface, especially at the bottom flow opening of the swirl filter, since the sediment content in the water here is high and the rotation speed is high. Therefore, the wear here is large, and the wall thickness of the underflow port portion of the cyclone filter needs to be regularly tested.
As shown in Fig. 8.17, during the maintenance of the swirl filter the following matters shoud be noted:

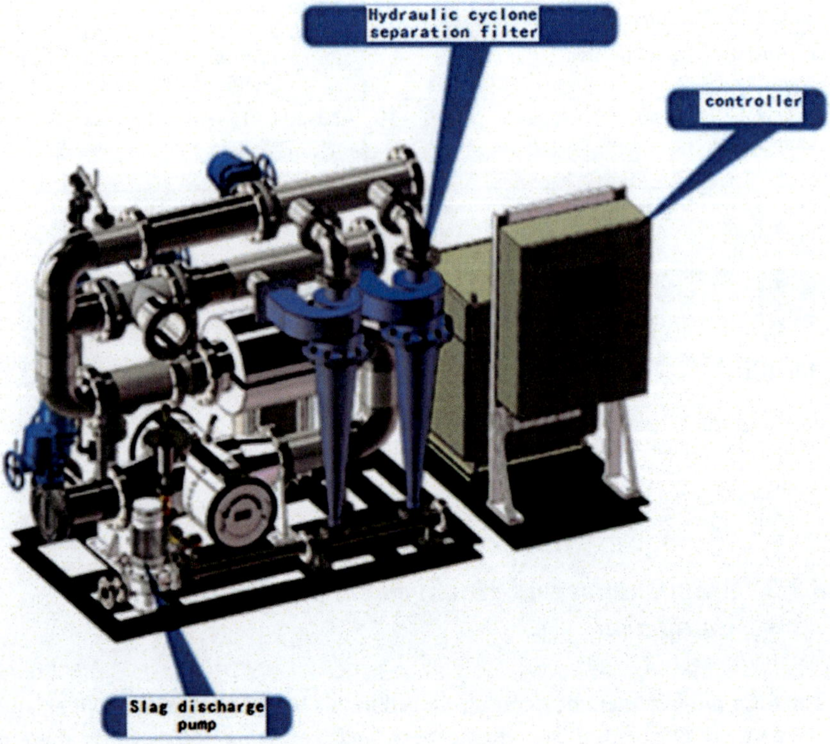

Fig. 8.17 Swirl filter in BWMS

① Turn on the power of the sewage pump, flush the drain pipe with clean water, and rinse the sewage and sludge at the underflow port.
② Close the inlet and outlet valves of the swirl filter and disconnect the system power supply.
③ Make sure that the swirl filter drains water before servicing.
④ Remove the bolts from the swirl inlet flange.
⑤ Remove the bolts from the overflow flange.
⑥ Remove the bolts from the bottom port flange.
⑦ Check the wear condition around the underflow port and replace if necessary.
⑧ Open the inlet and outlet valves of the swirl filter after reassembly, and turn on the system power.
⑨ Check for any leakage.

2 **The maintenance of Ultrasonic cavity purifier** [3]
The ultrasonic cavity purifier uses the cavitation effect of the ultrasonic wave to impact the surface of the cell membrane of the microorganisms in the ballast water, so that the cell membrane shrinks or ruptures under the action of high

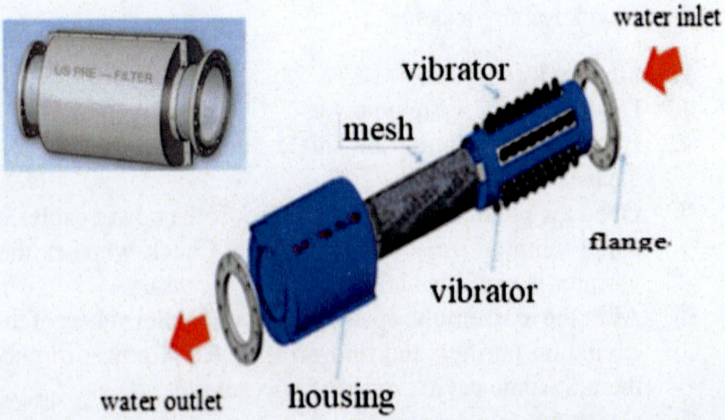

Fig. 8.18 Decomposition diagram of US cavitation purifier

pressure and mechanical shearing of the ultrasonic cavitation vacuole. As shown in Fig. 8.18, an ultrasonic vibrator is uniformly arranged on the four surfaces of the housing inside the protective cover, and a vibration transmitting mesh is installed in the ultrasonic cavity. Due to the cavitation effect of the ultrasonic wave, cavitation is easily formed in the ultrasonic cavity. For the woven vibration transmission net, because of the fine diameter of the stainless steel wire of the mesh SUS316L, it is easily broken under the action of cavitation and affects the efficiency of vibration transmission, and the connection where the ultrasonic cavity is connected to the inlet and outlet flanges is also rounded. Cavitation and strong ultrasonic vibration are easily broken, and the life of the ultrasonic vibrator is about 3000 h. Since the vibrator vibrates continuously on the cavity wall of the ultrasonic cavitation, the vibration of the vibrator power cord is loosened and the detachment also needs to be maintained. Therefore, the maintenance of the ultrasound cavitation purifier is divided into two types: maintenance of the cavity structure and maintenance of the vibrator.

(1) **The maintenance of cavity structure**

① Close the valves at both ends of the inlet and outlet of the ultrasonic cavitation purifier and disconnect the system power supply.

② Make sure that the ultrasonic cavitation purifier has drained water before servicing.

③ Remove the bolts on the inlet flange of the ultrasonic cavity purifier with proper support.

④ Under appropriate support, remove the bolts from the outlet flange of the ultrasonic cavity purifier.

⑤ Remove the ultrasonic cavitation purifier, check the damage of the inlet and outlet ports and the transmission network, and replace them if necessary.

⑥ After the reassembly, open the inlet and outlet valves of the ultrasonic cavitation purifier and turn on the system power supply.

⑦ Check for any leakage.

(2) **Vibrator maintenance**

① Disconnect the system power.

② Remove the bolts on the shield flange of the ultrasonic cavitation cleaner.

③ Open the protective cover and check if the power cables on the ultrasonic vibrator are loose or fall off. Check whether the ultrasonic vibrator is working and replace it if necessary.

④ After the reassembly, open the inlet and outlet valves of the ultrasonic cavitation purifier, and turn on the system power to check whether the ultrasonic cavitation purifier is started.

⑤ Check for any leakage.

3 **Ultrasonic cleaner maintenance**

Ultrasonic cleaners are divided into two categories in ballast water and marine bio-management systems. One is a columnar ultrasonic cleaner, which is used to install in multi-filter backwash filters. Ultrasonic cleaning is performed on filter elements that are rotated to nearby flushes as shown in Fig. 8.19. The other type is multi-vibrator planar cleaner, which is installed perpendicular to the ballast water flow direction and is installed on a UVC sterilizer without a mechanical descaler. The device for cleaning the quartz sheath of the UV lamp is shown in Fig. 8.20. The device is faced with a quartz sheath and works in conjunction with the ballast water biological management system to prevent scaling on the quartz sheath.

Fig. 8.19 Columnar US cleaner

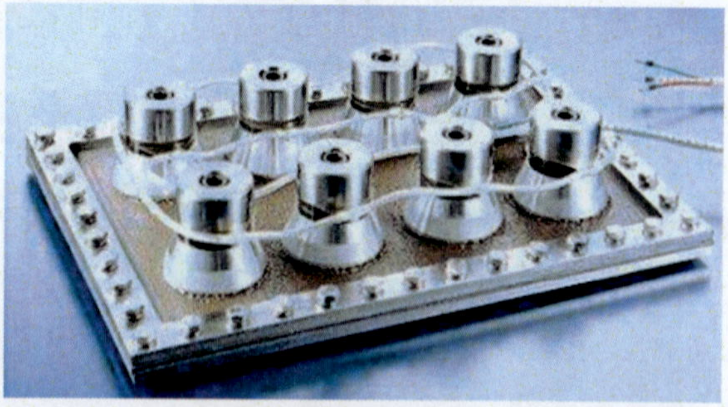

Fig. 8.20 Planar US cleaner

(1) **The maintenance of columnar ultrasonic cleaner** [4]
① Close the inlet and outlet valves of the multi-core backwash filter and disconnect the power supply.
② Make sure the water has been drained before servicing the multi-core backwash filter.
③ Remove the bolts on the cylindrical ultrasonic cleaner flange at the upper end of the multi-core backwash filter.
④ Remove the column ultrasonic cleaner and check if the seal rubber gasket is damaged and replace it if necessary.
⑤ After reassembling, open the multi-core backwash filter inlet and outlet valves and turn on the power.
⑥ Check for any leakage.

(2) **Maintenance of Planar Ultrasonic Cleaner Maintenance**
① Close the valves at both ends of the inlet and outlet of the UVC sterilizer without mechanical cleaning mechanism and disconnect the power supply.
② Make sure that the UVC sterilizer without mechanical cleaning mechanism has drained water before servicing.
③ Remove the bolts on the cover flange of the UVC sterilizer surface ultrasonic cleaner cover without mechanical cleaning mechanism.
④ Inspect the surface ultrasonic cleaner to check if the seal rubber gasket is broken and replace it if necessary.
⑤ After reassembly, open the valves at both ends of the UVC sterilizer without mechanical cleaning mechanism and turn on the power.
⑥ Check for any leakage.

4 **The maintenance of membrane filter** [5]
The membrane filter is a pre-treatment filter for inert gas sterilization system, with high filtration accuracy. Its structure and working principle are shown in Fig. 8.21.The ballast water enters from the filter inlet and is filtered by the membrane filter. When the membrane filter is blocked by impurities, the system closes the inlet and outlet valves, opens the drain valve and the compressed air valve, and injects high pressure air into the membrane filter, which will adhere

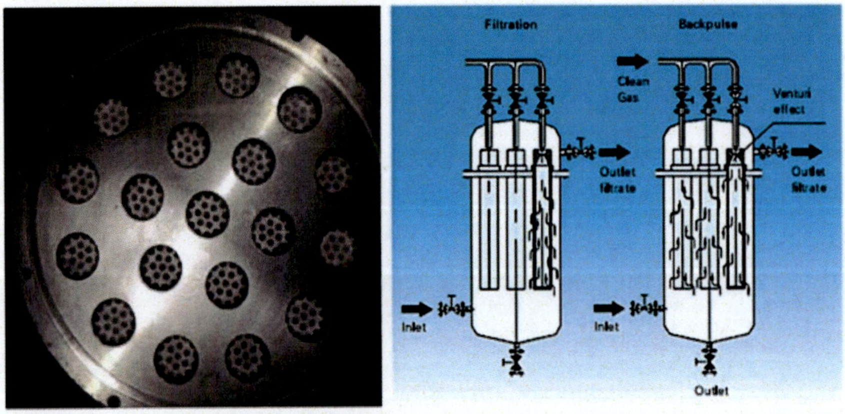

Fig. 8.21 Membrane filter element and principle

to the membrane filter. The impurities are blown off and discharged from the sewage pipe.

Due to the high precision of the membrane filter and the porosity of 10 μm, it is easy to clog when loading work in coastal waters or inland rivers with high sediment content. Therefore, regular maintenance is required. During maintenance, the following matters should be noted:

① Close the valves at both ends of the membrane filter inlet and outlet, and disconnect the power supply.

② Make sure the water is drained before the membrane filter is repaired.

③ Remove the high pressure air pipe and solenoid valve on the upper cover of the membrane filter

④ Remove the bolts from the upper flange of the membrane filter cover.

⑤ Remove the upper end cap of the membrane filter, drop the membrane filter assembly, and inspect the membrane filter.

⑥ Check the membrane filter element packing and the membrane filter seal. Whether the upper cover gasket is damaged or not, replace it if necessary.

⑦ After reassembling, open the inlet and outlet valves of the membrane filter and turn on the power.

⑧ Check for any leakage.

8.3 Maintenance and Management of Sterilization Unit

8.3.1 Maintenance and Management of the UVC Sterilization Unit

YP≋BWMS® medium-pressure ultraviolet disinfection device consists of the following three parts, as shown in Fig. 8.22, and its detailed component list is shown in Table 8.19 [1].

1 **Inlet and outlet component**
A picture of the YP≋BWMS® UV disinfection device is shown in Fig. 8.23 and its components are shown in Table 8.20.
According to the "YP≋BWMS UV disinfection device exploded view" shown, component #1 is the inlet and outlet component. When disassembling and installing the component of the inlet and outlet, the following matters should be noted::

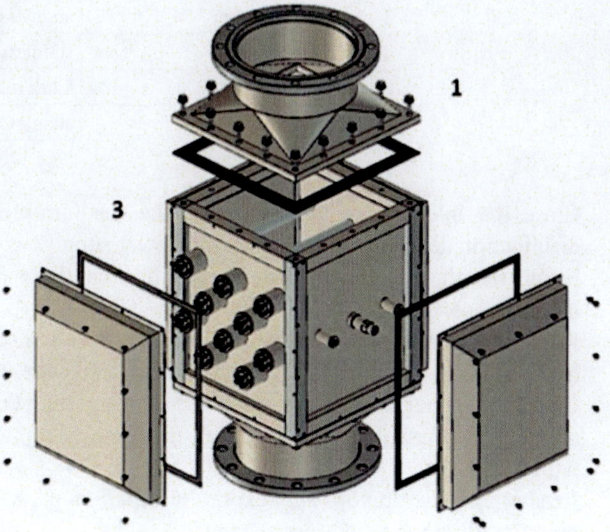

Fig. 8.22 YP≋BWMS® UVC sterilization device diagram

Table 8.19 YP≋BWMS® UVC sterilization device component details

No	Name	Materials	Quantity
1	Inlet and outlet component	Stainless steel	2
2	Chamber cover assembly	Stainless steel	4
3	UV lamp seal assembly	Stainless steel	10

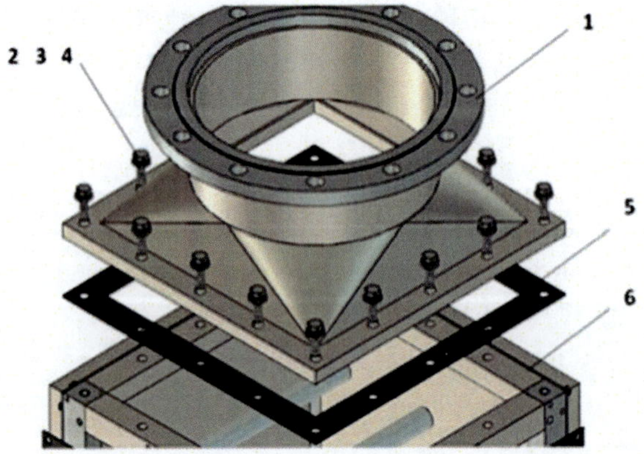

Fig. 8.23 Decomposition structure of inlet and outlet components of UVC sterilizing equipment

Table 8.20 UVC sterilizing equipment inlet/outlet component details

No	Name	Materials	Quantity
1	Inlet and outlet component	stainless steel	1
2	Flat gasket	stainless steel	16
3	Spring washer	stainless steel	16
4	Fastening bolt	stainless steel	16
5	Gasket	Cork rubber	1
6	Cavity	stainless steel	1

(1) Close the inlet and outlet valves of the medium-pressure ultraviolet disinfection device and disconnect the power supply.
(2) Ensure that the medium pressure UV sterilizer has drained the water before maintenance.
(3) Remove the bolts and washers 2, 3, 4 of the inlet and outlet components of the medium-pressure ultraviolet disinfection device.
(4) Check the damage of the gaskets of the inlet and outlet component of the medium-pressure ultraviolet sterilizer. If necessary, replace component #5 with a new one.
(5) Reassemble the import and export installation parts of the medium-pressure ultraviolet disinfection device.
(6) Open the inlet and outlet valves of the medium-pressure ultraviolet sterilizer and turn on the power.
(7) Check for any leakage.

In addition, during installation and maintenance, the following matters should be noted:

(1) During maintenance, the system shall be shut down, the inlet and outlet valves of the medium-pressure ultraviolet disinfection device shall be closed, and no water shall accumulate in the cavity of the device.

(2) In the disassembly process, remember the number of parts removed, visually inspected, and keep them properly.

2 **Chamber cover assembly**

The chamber cover assembly is shown in Figure 8.24. Its function is to protect the UV lamp assembly. The components are shown in Table 8.21.

According to the "YP≋BWMS UV disinfection device exploded view", component #2 is the chamber cover assembly. When it is necessary to repair the ultraviolet sterilizing assembly, the UV lamp sealing assembly, and the sensor, the chamber cover plate of the medium-pressure ultraviolet sterilizing device arere-moved in turn according to the "exploded view of the chamber cover assembly" and the various components are inspected and repaired.

(1) Close the inlet and outlet valves and disconnect the power supply.
(2) Ensure that the UV sterilizer has drained water before servicing.
(3) Disassemble the chamber cover bolts and gaskets 1,2,3, and 4 of the medium-pressure ultraviolet disinfection device one by one.

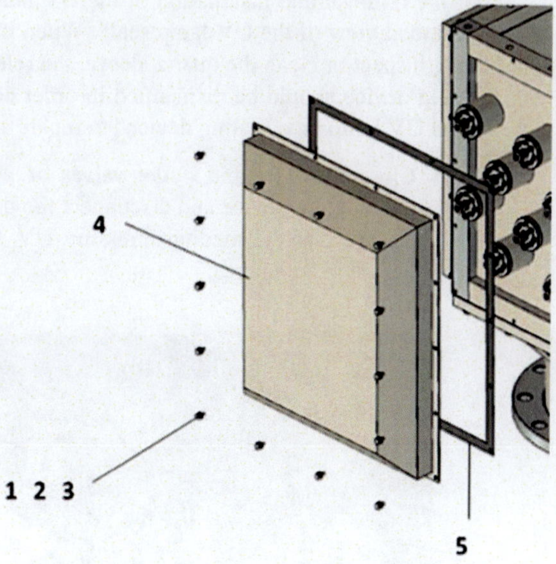

Fig. 8.24 Chamber cover assembly

Table 8.21 Details of the chamber cover assembly

No	Name	Materials	Quantity
1	Spring washer	stainless steel	14
2	Flat gasket	stainless steel	14
3	Fastening bolt	stainless steel	14
4	Chamber cover	stainless steel	1
5	Chamber cover gasket	Cork rubber	1

(4) Check the damage of the gasket of the cover assembly of the medium-pressure ultraviolet sterilizer. If necessary, replace it with a new one.
(5) Assembly of medium-pressure ultraviolet disinfection device chamber cover bolts and gaskets 1,2,3, and 4.

In addition, during installation maintenance, the following matters should be noted:

(1) Shut off during maintenance, close the inlet and outlet valves of the medium-pressure ultraviolet sterilizer, and ensure that there is no stagnant water in the ultraviolet sterilizer.
(2) When disassembling the chamber cover of the medium-pressure ultraviolet disinfection device, be aware of the residual temperature of the cover plate.
(3) During disassembly, visually inspect the removed parts and keep them properly.

3 **UV lamp seal assembly**

The UV lamp seal assembly is shown in Figure 8.25, and its detailed components are shown in Table 8.22.

For the removal and installation of the UV lamp, please proceed according to “exploded view of the UV lamp seal”. When it is necessary to replace the UV lamp, inspect or clean the quartz sleeve, the relevant parts of the UV lamp tube sealing device should be dismantled in order according to the “exploded view of the UV lamp tube sealing device” to repair the relevant parts.

(1) Close the inlet and outlet valves of the medium-pressure ultraviolet disinfection device and disconnect the power supply.
(2) Ensure that the medium pressure UV sterilizer has drained the water before maintenance.

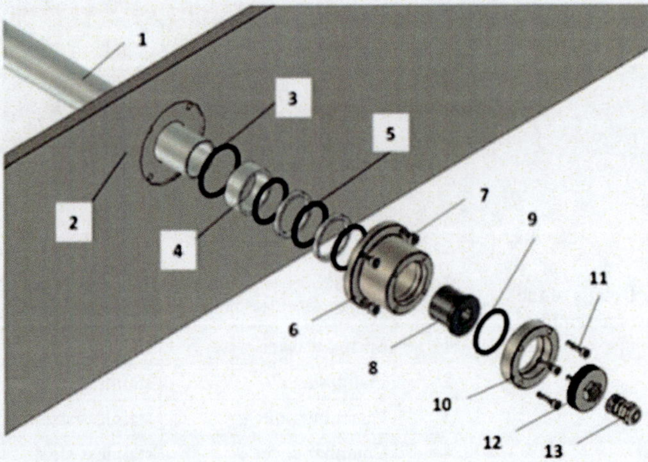

Fig. 8.25 UVC lamp sealing device decomposition diagram

Table 8.22 Details of UVC lamp sealing device

No	Name	Materials	Quantity
1	Quartz sleeve	quartz	1
2	Medium pressure ultraviolet disinfection device chamber	stainless steel	1
3	"O" ring	EPDM rubber	1
4	Sealing bush	Teflon	3
5	"O" ring	EPDM rubber	4
6	Lamp outlet terminal base	stainless steel	1
7	Fastening bolt	stainless steel	4
8	Lamp bushing	Teflon	1
9	"O" ring	EPDM rubber	1
10	Lamp outlet cap	stainless steel	1
11	Fastening bolt	stainless steel	4
12	Cable gland base	stainless steel	1
13	Cable glands	Copper alloy	1

(3) Detach the cables of the UV lamp to be maintained and mark it.
(4) Remove the cable glands and the cable glands 1,3,1,2.
(5) Take out the UV lamp bush 8.
(6) Carefully remove the UV lamp.
(7) Remove the lamp outlet cap and save it properly.
(8) Remove the lamp outlet terminal base 6.
(9) On the side opposite to the UV lamp, remove the lamp end base. At this point, the quartz sleeve sealing device was disassembled and the quartz sleeve can becarefully removed.
(10) Check the condition of the quartz tube, UV lamp, and seal, and clean or renew if necessary.
(11) Put the "O" ring and seal bushing on the quartz sleeve 3, 4.
(12) Replace the lamp outlet terminal base and lamp terminal base.
(13) Replace the seal and lamp holder cover in sequence.
(14) Test pressure on the UV lamp seals and ensure that there is no leakage.
(15) Ensure that there is no leakage of the UV lamp sealing device, and connect and fasten the UV lamp cable and the cable gland.

In addition, when the ballast water management system is in operation, the UV lamp tube needs to be replaced and installed on-line due to the damage of the UV lamp, that is, when the UV sterilization device is not closed or stopped, the malfunctioning UV lamp tube is maintained. The online installation and replacement steps should refer to the description of steps (1) to (6) in "Demolition and Installation of UV Lamp Seals".

In addition, during installation and maintenance, the following matters should be noted::

(1) When the UV lamp sealing device is overhauled, the system should be shut down, the inlet and outlet valves of the medium-pressure ultraviolet disinfection device should be closed, and there should be no accumulation of water in the medium-pressure ultraviolet disinfection device.
(2) During disassembly, visually inspect the removed parts and keep them properly.
(3) Please wait for a moment (about 3 to 5 min) before the UV lamp sealing device is overhauled. After the parts are cooled, perform the disassembly and assembly operations again.
(4) Quartz sleeves, UV lamps, and ceramic bushings are fragile items. Special care must be taken during their removal, replacement, and installation and maintenance. After being disassembled, it should be properly preserved so as not to cause personal injury and damage to parts and components.
(5) The quartz sleeve cleaning fluid may cause discomfort to the personnel. When manually cleaning the quartz sleeve, please make necessary protective measures. It should be conducted in a ventilated place or outdoors, and pay attention to the fireproof and explosion protection of the cleaning place.
(6) Keep the surface of the UV lamp clean and do not touch the light emitting section of the UV lamp with your hands.
(7) After the UV lamp sealing device is reinstalled, pay attention to the sealing of the quartz sleeve end face in the sealing device. And keep the device surface dry.

4 **UV lamp life**

The normal service life of the UV lamp inthe ballast water management systemYP≈BWMS® is about 1500 h. The service life of UV lamps varies according to their operating conditions (eg, start-stop frequency, cooling conditions, etc.). Therefore, they must be monitored at any time to ensure the normal operation of the UV lamps.

5 **Quartz sleeve cleaning**

Quartz sleeves are an indispensable accessory in UV disinfection devices. Its role is to protect the UV lamp from the impact of water, while ensuring the transmittance of the UV lamp.Quartz sleeves for ballast water management systemYP≈BWMS® feature high purity and high light transmission (more than 90%).

Even though the ballast water management system YP≈BWMS® has a quartz sleeve automatic cleaning device, after the medium-pressure ultraviolet disinfection device works for 6 months (or other time) depending on the water quality condition, the quartz sleeve needs to be removed for cleaning. The cleaning steps are as follows:

(1) Turn off the power supply of the medium-pressure ultraviolet sterilizer and hang up a warning sign.

(2) Discharge the medium-pressure ultraviolet sterilizer from seawater, and confirm that the inlet and outlet valves are closed.
(3) Carefully remove the quartz sleeve to be cleaned and observe the condition of the quartz sleeve. If the quartz tube is damaged, it should be replaced.
(4) Clean the quartz sleeve with a sponge or a cotton cloth.

In addition, during installation maintenance, the following matters should be noted:

(1) Read the safety precautions of this manual again and take protective measures.
(2) The quartz sleeve cleaning fluid may damage the skin and cause discomfort to the personnel. Please perform protective work when cleaning.
(3) Quartz sleeve cleaning must be performed in a ventilated place without any direct or potential danger such as fire or explosion.
(4) Quartz sleeve cleaning fluid has a pungent odor. If you feel obvious discomfort or other adverse reactions, immediately stop the cleaning operation. Store the cleaning fluid and keep it properly. Bring uncomfortable personnel to ventilation and seek medical attention if necessary.
(5) Clean the cleaned quartz sleeve surface and clean it properly.
(6) For more information about quartz sleeve cleaning, contact the equipment manufacturer or agent

6 **Power distribution and control**

Good and stable working status of the power distribution and control system is a key factor for the normal operation of the equipment. Attention should be paid to the following mattersin the daily maintenance work of maintenance personnel:

(1) The humidity of the environment surrounding the power distribution cabinet and control cabinet must not exceed 90%. Pay attention to dust protection.
(2) Do not splash oily liquids or corrosive liquids on the touch panel surface of the control cabinet.
(3) Non-professionals are strictly forbidden to disassemble the circuit and electrical components.
(4) It is forbidden to place other sundries in the power distribution cabinet and control cabinet.
(5) Conduct dust removal and electrical line inspections at least once a year for distribution cabinets and control cabinets.
(6) Regularly check the internal wiring of the power distribution and control system.

8.3.2 Maintenance and Management of Other Types of Sterilization Units

In addition to the above-mentioned UV sterilizing units, the ballast water and marine biological management system also uses a variety of sterilization methods, such as photocatalytic sterilizing units, electrolysis or pharmaceutical sterilizing units, and inert gas sterilizing units.

1 **The maintenance of photocatalytic sterilization unit**
The structure of the photocatalytic sterilizing unit is very similar to that of the ultraviolet sterilizing unit. The difference lies in that the mechanical cleaning mechanism and the ultrasonic cleaning device are eliminated in the sterilization chamber, and a metal mesh structure with nano TiO2 catalyst attached is added around the UV lamp tube. Under ultraviolet light irradiation, in addition to ultraviolet radiation sterilization, the nano-TiO2 catalyst also reacts with water under the excitation of ultraviolet light to produce OH-, which oxidizes the cell membrane of the microorganism and renders the cell inactive. Therefore, for the photocatalytic sterilizing unit, the installation and maintenance method may refer to the installation and maintenance methods of the ultraviolet sterilizing unit.

2 **The maintenance of electrolysis sterilization unit**
The electrolytic sterilizing unit is a device that uses direct current to electrolyze seawater to produce a strong oxidant, sodium hypochlorite, and kills microorganisms in the seawater. Therefore, for its maintenance, the following matters need to be considered:

(1) Before the system is maintained, the fan should be turned on to remove hydrogen and chlorine that may remain in the swirl separator.
(2) Turn off the power, close the inlet and outlet valves of the electrolytic sterilizer, and ensure that there is no stagnant water in the electrolytic sterilizer.
(3) During disassembly, visually inspect the removed parts and keep them properly.
(4) Check the loss of the motor in the tube plate type electrolyzer and replace it if necessary; check the installation of the power lead on the electrode and replace it if necessary.
(5) When removing the tube plate type electrolyzer, waterproof rubber gloves shall be provided to prevent sodium hypochlorite from corroding human skin. Eye protection glasses and gas masks shall be provided to prevent harmful gas or corrosive liquid from splashing on the maintainer.
(6) Check the integrity of the sealing element and replace it if necessary.
(7) Check the differential pressure switch of the wind pressure of the fan and replace it if necessary.
(8) Check the hydrogen and chlorine leak sensors in the system and replace them if necessary.

Fig. 8.26 Method liquid medicine reservoir tank

(9) After the tube plate type electrolytic cell is reassembled, pay attention to the sealing of the device's face. And keep the device surface dry.

3 **The maintenance of Pharmacy Sterilization Unit** [6]
Since the pharmacy method is divided into liquid and solid, the maintenance and management of the system are different in the addition and preparation of the medicament, as shown in Fig. 8.26 and 8.27.
From the point of view of the two methods for storing liquids or powders with the pharmacy method, in addition to the main points of attention mentioned for the electrolytic sterilization methods (2), (3), (5), (6), and (9), there is, in addition, the maintenance work for adding the chemicals. For liquid drug addition, pay attention to the protection of the drug handling and the filling of the drug. The handling of the drug is shown in Fig. 8.28. The protection method is shown in Table 8.23. The filling process of the drug is shown in Fig. 8.29. The protection method is shown in Table 8.23 (Table 8.24).
The pharmacy method powdered drug handling process and protection method are the same as the method for transporting and protecting the liquid drug's deck. The process of handling it in the engine room is shown in Fig. 8.30. The loading process is slightly different depending on the structural platform of the ship the system is installed on, as shown in Fig. 8.31a, b.
As can be seen from Fig. 8.31, when the powdered medicament is added to the pharmaceutical filling station in the same deck space, the clearance between the

Fig. 8.27 Pharmaceutical storage tank

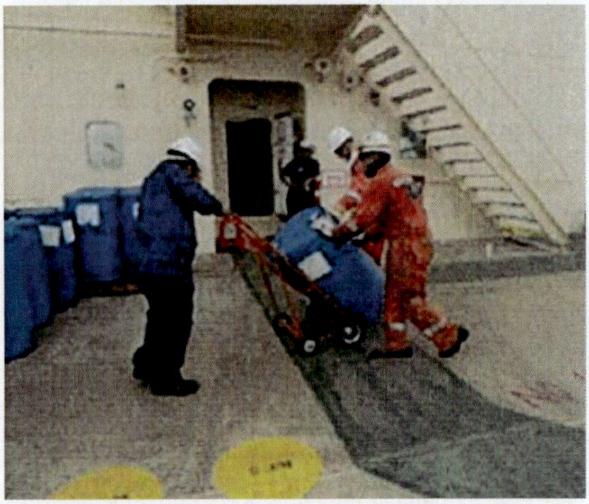

Fig. 8.28 Liquid disinfectant deck handling

Table 8.23 Precautions for deck handling of disinfectant

Personal protective wear	Personal protective wear instructions	Major equipment	Major equipment protection function
	When carrying drugs on the deck, the crew should wear safety helmets, goggles, gas masks, long-sleeved rubber non-slip gloves, and long boots. Prevent the erosion of medicaments		Hard hat: Prevent head injuries
			Goggles: Prevent liquid from splashing into eyes
			Gas mask: gas that prevents the pungent odor of the respiratory tract
			Long Sleeve Non-slip Rubber Gloves: Prevent Arms from Touching and Transporting
			Long boots: anti-slip and prevent splashing liquid on the feet

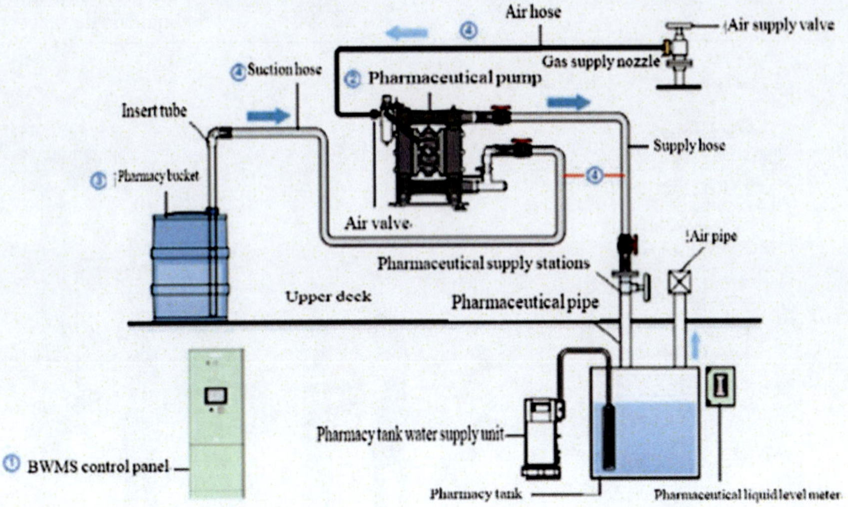

Fig. 8.29 Liquid disinfectant filling process

deck floors should not be less than 4 m; when the medicine on the upperdeck is filled onto the nextdeck, the headroom between the floor of the filling station and the upper deck should not be less than 3 m.

Table 8.24 Precautions for personnel protection when adding disinfectant solution

Personal protective wear	Personal protective wear instructions	Major equipment	Major equipment protection function
	The crew must wear waterproof, spatter-proof, non-slip,anti-toxic odor equipment throughout the maintenance of the drug system. Prevent the damage of the chemical to the crew and protect the crew's head in the cabin		Hard hat: Prevent head injuries
			Mask: Prevent liquid splashing on the eyes and face
			Gas mask: gas that prevents the pungent odor of the respiratory tract
			Raincoat: protect the body of the crew to prevent splashing
			Long Sleeve Non-slip Rubber Gloves: Prevent Arms from Touching and Transporting
			Long boots: anti-slip and prevent splashing liquid on the feet

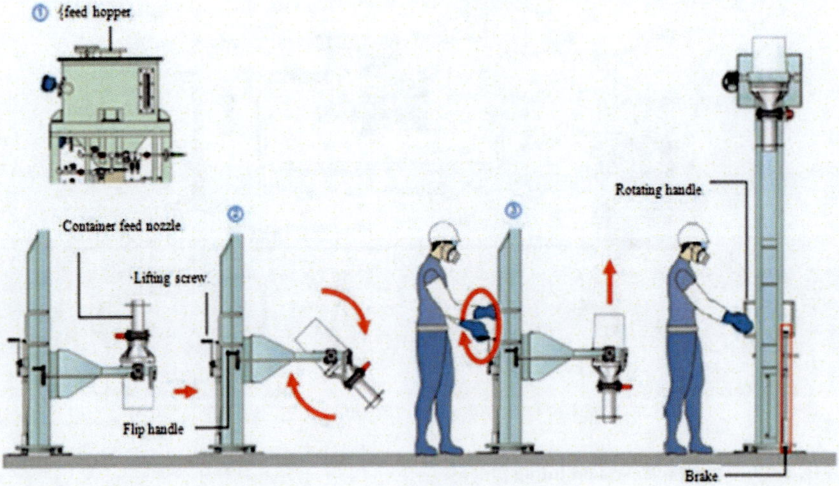

Fig. 8.30 Loading and filling process of powdered disinfectant cabin

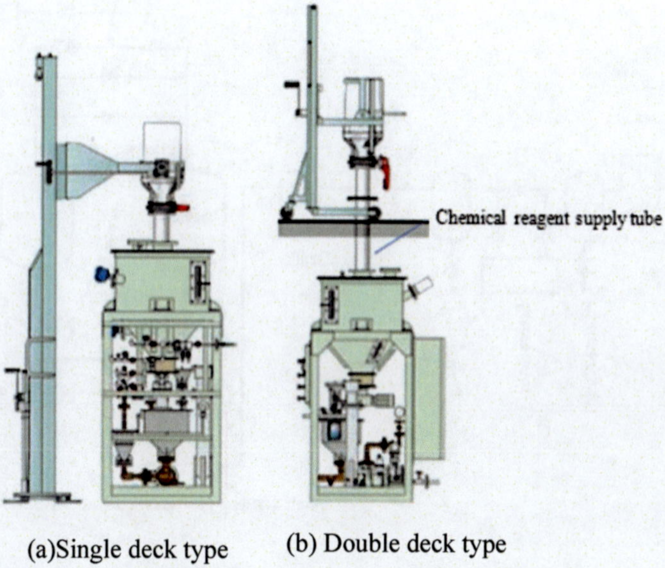

Fig. 8.31 Equipment in different types of filling

4 **Maintenance of inert gas sterilization system** [5, 7]
The main equipment of the inert gas sterilization system is the adsorption nitrogen-generating equipment through molecular sieves. Nitrogen is produced and injected into the ballast tanks to remove oxygen, so as to achieve the purpose of killing and inhibiting the growth of bacteria. The working principle of its equipment system is shown in Fig. 8.32.
Molecular sieve adsorber plays an auxiliary role in oxygen production and nitrogen production equipment. Although it is not the main equipment, it played a decisive role. Although its structure and operation are not complicated, it can cause serious consequences if it malfunctions. The most common faults are as follows:

(1) **Leakage of molecular sieve adsorber**
The adsorbent in the molecular sieve adsorber is a molecular sieve. The molecular sieve adsorbs moisture while also adsorbing other substances such as CO_2. At room temperature, the molecular sieve adsorber has a high purification effect on the purified air. The content of water vapor in the adsorbed air reaches a dew point of −70 °C. However, its surface area is small. After a certain amount of air is treated, the surface quickly condenses to saturation and cannot be resorbed. It needs to be heated and regenerated. If the molecular sieve adsorber leaks badly, it will accelerate the absorption of the final moisture, and the molecular sieve failure will accelerate, causing the pipeline to freeze in advance and the startup to

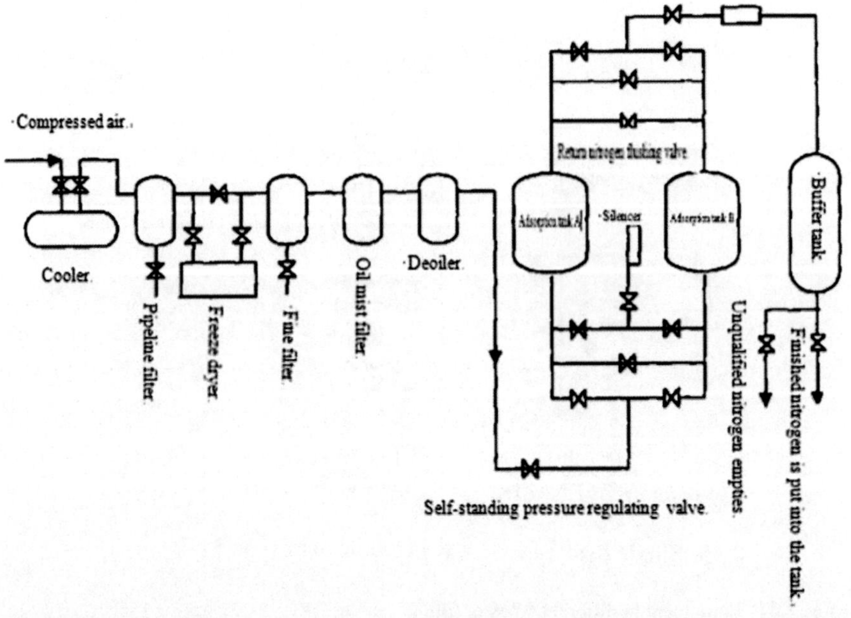

Fig. 8.32 Working principle diagram of adsorption nitrogen making equipment

fail. Therefore, the following must be noted in the leakage maintenance of molecular sieve adsorber:

(1) During operation of the system, attention should be paid to the fact that the flange bolts of the top cover of the molecular sieve adsorber are not tightened. If they are not tightened, the power supply must be turned off, the gaskets must be replaced, and the bolts must be tightened so that they do not leak.

(2) Check whether the gasket of the resistance thermometer socket is sealed (the platinum resistance is not tightened or the sealing gasket is defective and not sealed). If it is not sealed, replace the platinum gasket here.

(3) Check the molecular sieve adsorber tube rupture.If there is leakage, you should turn off the power supply, repair the cracks with argon arc welding. When themolecular sieve adsorber tube has wide and/or lengthy cracks, you should replace the molecular sieve adsorber.

(2) **Molecular sieve adsorber bursts when heating**

When the adsorber is working, the two drain valves of the water jacket should be opened so that the internal water can flow out. At the same time, the inlet and outlet valves are closed. During heating, the temperature of the electric furnace is about 300 °C. If the water in the jacket cannot flow out, the molecular sieve adsorber will heat the cooling water in the water jacket when it is working. Water evaporates at 100 °C and expands rapidly. If there is no pressure relief point, the pressure will become higher and higher. At the same time, as the temperature rises, the strength of the aluminum material also decreases.The air pressure of the intake air and the water pressure of the expansion act on the metal wall, and once it exceeds the strength of the aluminum material, it will burst. Therefore, when the adsorber is started, it is necessary to open the inlet valve and check whether the two discharge valves of the adsorber water jacket have water flowing out. When there is water flowing out, the inlet valve needs to be closed. If no water flows out, the two drain valves of the adsorber water jacket should be considered closed or blocked. If it is found that the two discharge valves of the adsorber water jacket are closed, they should be opened immediately; otherwise, the water inlet pipe may be blocked or the drain valve may be blocked.

(3) **Evacuation mouth with molecular sieve chips or powder blowing out**
The molecular sieve adsorber suction vacuum exhausts clean high-temperature gas during operation. If debris or white powder blows out at this time, it should be disposed of in time. Otherwise it will reduce the adsorption effect. When the chips or white powder are blown out, it should be judged that the molecular sieves are used for too long, the broken particles are increased, the delaminated powder is increased, the molecular sieve is too small, and the air flow causes the molecular sieve to continuously flip and rub the friction during work, and then, the filter glass wool is consumed and it lose the functionality offiltration. You need to open the molecular sieve adsorber cover to remove the filter layer, remove the bottom plate, use a clean oil-free large basin to hold the molecular sieve, take a sieve with a sieve equivalent to approximately 1/3 of the diameter of the sieve to screen the molecular sieve, continue to use the remaining one, restorethe molecular sieve adsorber and add a new one to the bottom of the well plate; add enough glass wool when it is missing. The casing on the exhaust pipe should be equipped with a filter to prevent particles and impurities from entering the pipeline and the valve, causing blockage of the pipeline and the opening and closing of the valve. If it enters the tower, it can also cause poor distillation of the fractionation tower.

(4) **Molecular sieve adsorber leaks outward**
When the temperature of the enclosure is high during heating, the outward leakage of the molecular sieve adsorber generally occurs at the outlet joint of the heating furnace. This is followed by heating the air into the joint of the molecular sieve adsorber valve or the rupture of the line between the two locations. These parts are often subject to natural cooling after high temperatures repeatedly, and due to the vibration of the compressor operation, the joints are loosened. This section of the pipeline is at a high temperature due to heating, and the pure copper tube is softened and peeled off. The tube wall becomes thinner, and because of the high temperature,

the softening strength of the pure copper tube decreases, and the air pressure cannot expand at the thinnest part. There are two reasons for the high shell temperature during heating. One is that the pipeline ruptures and the air leakage from the joint is heated by the heated gas. Second, there are too few thermal insulation materials in the molecular sieve adsorber. The space gas in the tank is heated by the heat pipe and the electric furnace. At this time, brazing was used to repair the broken pipe, and the leaking nut was tightened to add a thermal insulation material to the molecular sieve adsorber lacking the thermal insulation material.

8.4 Maintenance of Other Important Components in Ballast Water Management Systems

In addition to the filter unit and sterilizing unit, the ballast water management system also includes an electric valve group, an electromagnetic flowmeter, a functional pump, and a power supply unit that constitute the system. These important components control the loading and unloading process of the ballast water management system and the emergency discharge operation process. This book will use the Ballast Water Management System YP≋BWMS® developed by Jiangsu University of Science and Technology as an example to introduce these important components [1].

8.4.1 The Maintenance and Management of Electric Butterfly Valve

The pipeline of the ballast water management system YP≋BWMS® is equipped with an electric butterfly valve controlled by a PLC to regulate the direction and size of the water flow. The control system controls the opening time and opening angle of the electric butterfly valve according to the operating conditions of the equipment. Therefore, the following tasksshould be completed before the installation of the electric butterfly valve:

(1) Carefully check that the valve usage is in accordance with its performance specifications.
(2) Check the valve passage and sealing surface for debris and remove it in time.
(3) The valve should be installed in time after unpacking. Do not loosen any fastening screws on the valve.
(4) Hard seal, soft seal with standard flange.
(5) Before installation, the commissioner or installer should install in strict accordance with the wiring diagram.

Ballast water management systemsYP≋BWMS® are equipped with "passive contact" electric butterfly valves and "smart regulator" electric butterfly valves. The "passive contact type" electric butterfly valve is installed in the inlet and outlet of the self-cleaning filter, the self-cleaning filter bypass line and the system return line. When the electric butterfly valve actuator receives an appropriate command, the actuator drives the valve to the fullyopen and fully closed position. The "smart adjustable" electric butterfly valve is installed in the system to the ship's ballast tank pipeline. The electric butterfly valve actuator drives the valve to a suitable opening position based on the variation of the controlled variable (flow, pressure, etc.) in the pipeline. Passive contact type electric butterfly valve control circuit is shown in Fig. 8.33. Intelligent adjustment type electric butterfly valve control circuit is shown in Fig. 8.34.

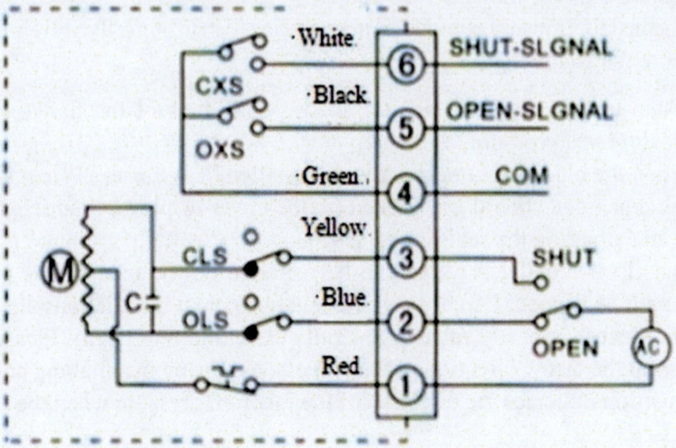

Fig. 8.33 Passive contact electric butterfly valve control circuit diagram

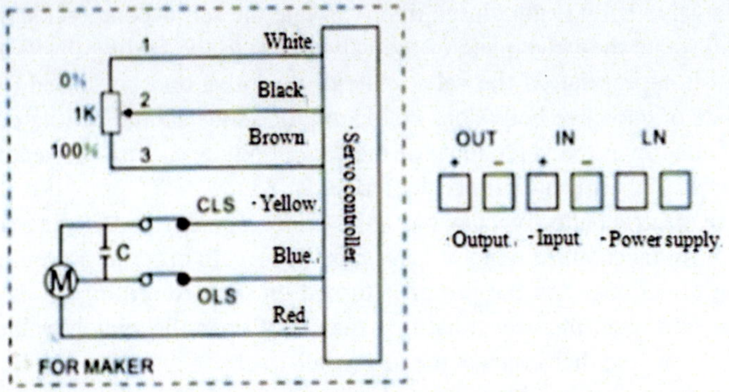

Fig. 8.34 Intelligent control electric butterfly valve control circuit diagram

The opening and closing of the valve is achieved through a switching circuit and it outputs a set of passive position signals indicating that the valve is fully open and fully closed.

The standard signal input by an external computer or industrial instrument controls the opening and closing angle of the valve and synchronously feeds back the standard signal corresponding to the output.

It is worth noting that the dotted line in the control circuit diagram is the internal circuit of the electric device, and the part other than the dotted line is only for reference when wiring.

At the same time, special attention should be paid to the installation of the electric butterfly valve. It is not possible to connect the power lines of two or more electric devices in parallel, and it is not possible to use the same connection point to control several electric devices. Otherwise, the electric butterfly valve may become out of control and the motor may be overheated.

For the installation and commissioning of the electric butterfly valve in the system, please note the following:

(1) When testing pressure, be sure to use both ends of the flange to clamp or pressure test pressure test bench.
(2) When the electric butterfly valve is installed, it is closed. When welding, the valve passage should be blocked by the cover to prevent particles and debris from damaging the sealing surface.
(3) The electric butterfly valve can be installed on the pipe at any angle, but it cannot be flipped. Usually vertical and horizontal installations are used.
(4) The electric butterfly valve is generally used unidirectionally. Please pay attention to the arrow direction of the valve body during installation, and the arrow direction indicates the direction of the medium pressure when the valve seal is closed.
(5) Although the product undergoes strict inspection and pressure test before leaving the factory, there are also individual product vibration screws that change during transportation and need to be readjusted. If the valve plate of the valve body is not closed tightly, rotate the screw counterclockwise a little and rotate the valve plate of the valve body to the sealing position. Lock the nut if appropriate.If the valve plate of the valve body is closed (or the valve plate of the valve body does not open), the screw should be rotated clockwise a little, open the valve plate of the valve body and close the sealing position again. Lock this nut properly afterwards.
(6) The electric butterfly valve has adjusted the opening and closing stroke of the control mechanism when it leaves the factory, in order to prevent mistakes in the direction when the power is turned on. Before turning on the power for the first time, the user must first manually open the switch to the half-open position, and then connect the jog switch to check that the indication direction is the same as the opening direction.
(7) Each electric butterfly valve (including sewage solenoid valve) of the ballast water management systemYP≈BWMS® comes with a handle randomly

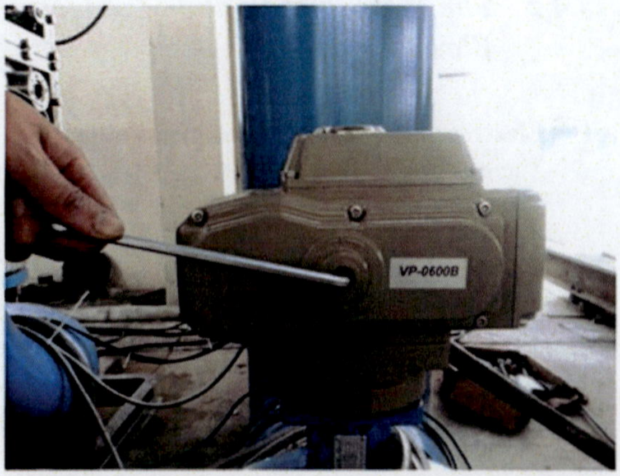

Fig. 8.35 Manual operation of electric butterfly valve schematic

shipped.When the electric butterfly valve cannot be automatically opened and closed, the handle can be inserted into the manual operation port to drive the valve, as shown in Fig. 8.35. When not in use, please remove the handle and store it separately, and then fix the rubber dust plug screw.

For the maintenance of the electric butterfly valve in the system, you should pay attention to the following matters:

(1) The valve plate and the seal ring of the butterfly valve have impurities, and impurities should be removed to clean the valve cavity.
(2) The valve plate and the closed position of the butterfly valve of are inconsistent, and the limit screws of the actuator such as the worm gear or the electric actuator should be adjusted so as to achieve the correct valve closing position.
(3) When the flange bolts of the outlet are not uniformly loaded or not pressed, the flange plane and the bolt pressing force shall be checked and pressed evenly.
(4) The direction of pressure test is not as required, and spinning should be performed in the direction of the arrow.

The failure of the butterfly valve is mainly manifested in the leakage of both ends of the valve, and its possible failure phenomena are as follows.

(1) The gasket on both sides fails.
(2) The pipe flanges are not evenly pressed or pressed tightly.

In response to the above-mentioned failure phenomena, the main solutions are:

(1) Replace the gasket.
(2) Tighten the flange bolts (uniform force).
(3) Remove the pressure ring of the lower valve and replace the seal ring and the failure gasket.

8.4.2 *The Maintenance and Management of Electromagnetic Flowmeter*

1 **Electromagnetic flowmeter installation**

The electromagnetic flowmeter plays a role in the ballast water management system to monitor the amount of water entering and leaving the ballast tank during loading and unloading of the ballast water and to monitor the dose of sodium hypochlorite for UV metering, electrolysis, or medicament supply of the sterilizing unit. The water-carrying management system is a system-level monitoring sensor. As the ballast water management system is a IMO regulatory product, the entry into force of this regulation will be traced back to many ships currently in operation, and therefore involves the refitting of a large number of ships. Therefore, the product development companies of ballast water management systems have taken into account the future ships. In the modification, each functional unit was modularly designed to install the system on a converted vessel. In order to be able to effectively monitor the inflow and outflow of ballast water, the flowmeter installed as shown in Fig. 7.2 should be filled with liquid. The straight section of the pipe, such as the vertical section of the pipe or the horizontal section filled with liquid, meets the measurement. In the design of the position of the flowmeter modified by the ship ballast water system, attention should also be paid to the following matters [1, 8]:

(1) It should be installed at the rise of the pipe, as shown in Fig. 8.36.
(2) The sensor must not be installed at the inlet of the pump. It should be installed at the outlet of the pump, as shown in Fig. 8.37.
(3) The measuring electrode's axis must be approximately horizontal, as shown in Fig. 8.38.
(4) The control valve and shut-off valve should be installed downstream of the sensor and should not be installed upstream of the sensor, as shown in Fig. 8.39.

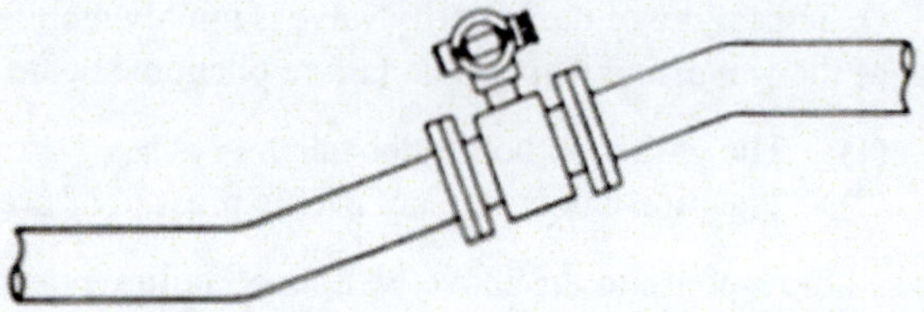

Fig. 8.36 Installation example 2 of flow sensor

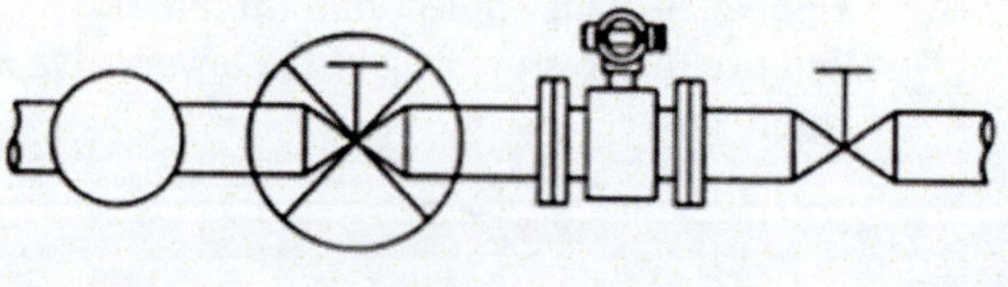

Fig. 8.37 Installation example 3 of flow sensor

Fig. 8.38 Measuring electrode mounting

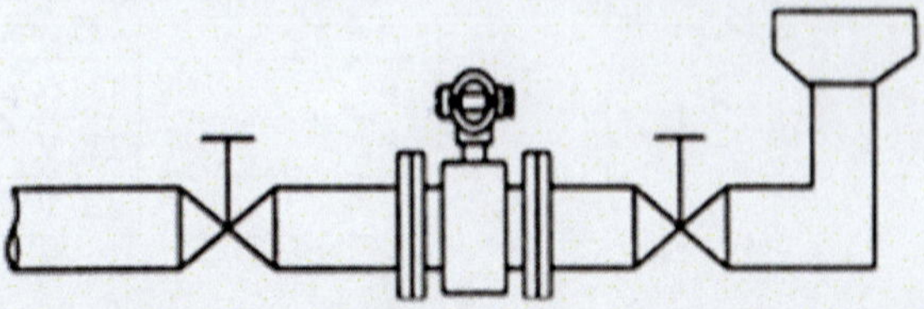

Fig. 8.39 Installation example 4 of flow sensor

(5) When installed on open discharge pipelines, it should be installed in the lower part of the pipeline, as shown in Fig. 8.40.

2 **Electromagnetic flowmeter terminal description**

Converter and sensor connection methods are shown in Table 8.25.

When the split type converter and the sensor are used together, if the conductivity of the fluid to be measured is greater than 50μS/cm, the type of the RVVP2 × 32/0.2 PVC, PVC sheathed metal mesh shielded signal cablemay be used. It should not exceed 100 m. Double-core double-shielded signal cable with

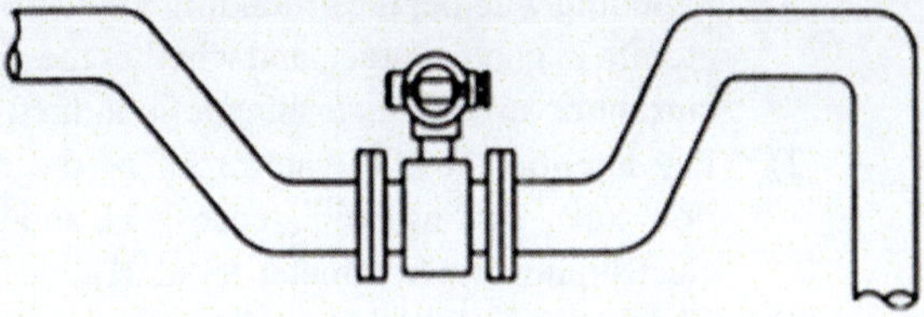

Fig. 8.40 Installation example 5 of flow sensor

Table 8.25 Converter and sensor connection method

Signal line	Signal line description
SIG +	Positive sensor signal
SIG–	Negative sensor signal
SGND	Sensor common

Table 8.26 Excitation wirewiring diagram

Excitation line	Excitation line description
EXT1	Sensor excitation terminal
EXT2	Sensor excitation terminal

Table 8.27 Statement of wiring of excitation cable and signal cable

Terminal symbol	Terminal description
IOUT:	Current output (active)
ICOM:	Current output (active)
PUL:	Bidirectional flow frequency (pulse) output
COM:	Ground for frequency (pulse) alarm output
ALML:	Lower limit alarm output
ALMH:	High limit alarm output
A + :	485 communication input (optional function)
B−:	485 communication input (optional function)
N:	220 V power input
L:	220 V power input

equipotential shielding can be used when the measured conductivity is less than or transmitted over long distances. The wiring of the excitation wire is shown in Table 8.26.

Excitation current adopts two-core insulated rubber flexible cable. The recommended model is YHZ-2 × 1 mm^2. The length of the excitation current line is the same as that of the signal cable. When using the STT3200 dedicated cable, the excitation cable and the signal cable are combined into one. The wiring instructions are shown in Table 8.27.

3 Daily maintenance and management of electromagnetic flowmeter

(1) Always check the grounding system of the electromagnetic flowmeter, including whether the grounding wire, grounding ring, grounding terminal, etc. are in good contact, and whether the grounding system will have strong current or strong electromagnetic induction.
(2) Pay attention to the waterproof of the instrument: wiring port, protection tube, sealing joint, case seal and so on.Check the lining of the electromagnetic flowmeter regularly.
(3) Check the installation of the pipeline frequently to see if there is any leakage, vibration of the pipeline, and loose bolts, etc.
(4) Check the flowmeter's preservative insulation at regular intervals.
(5) Measure whether seawater impurities contaminate the electrode or precipitate and foul in the wall of the measurement tube. It should be periodically cleaned.

8.4.3 The Maintenance and Management of High Pressure Cleaning Pump and Sewage Pump

The high-pressure cleaning pump and the sewage pump of the ballast water management system YP≋BWMS® are centrifugal pumps. Its installation and maintenance requirements are as follows.

1 **High-pressure cleaning pump and sewage pump installation.**

(1) The anchor bolts must be properly installed and fastened, and no loose or loose conditions are allowed.
(2) According to the type of fluid, select the appropriate seal.
(3) The pump inlet and outlet must be installed horizontally.
(4) All piping and fittings connected to the pump body should be firmly installed without looseness or leakage.

2 **The trial operation of high-pressure cleaning pump and sewage pump shall meet the following requirements**

(1) The steering of the driving motor should be the same as that of the pump.
(2) There should be no looseness at each fixed joint, and lubricants should be added to each lubrication part.
(3) The rolling stock should be flexible without any abnormalities.

3 **The maintenance of high pressure cleaning pump and sewage pump**

(1) The alignment of the driver and the pump should be readjusted if there is a large difference from the original data.
(2) Dismantle and inspect the rotor, shaft, and bearing wear of the pump and carry out nondestructive testing.
(3) Perform macro inspections and inspections on the components of the pump.
(4) Check the wear of the pump, eliminate the wear gap, and increase the efficiency of the pump.
(5) Regularly check the seal of the pump if necessary.
(6) Clean and purge dirt from the pump.
(7) Eliminate leakage of pumps and auxiliary parts and check the oil system.

8.4.4 The Maintenance and Management of Power Unit

The power supply unit is a device that provides the power supply voltage for the ultraviolet ballast. The technical parameters are: three-phase alternating current, 380–480 V, + 6/–10%, and 48–62 Hz [9].

Through the dedicated computer and software, the maximum output power of the UV ballast can be arbitrarily controlled and set, and the infinite power control from 10 to 100% can be performed at the same time.

When installing and wiring the power supply unit, it must be done by an authorized professional. Prior to any work on the ballast or lamp, the power must be disconnected! Even if the power is turned off, the deadly residual voltage may still exist for 10 min! Never start any ballast that is obviously damaged! Do not open the ballast! You must follow the guidelines in the user manual. Improper operation can bring danger and it can also result in loss of product warranty.

1 **Installation and erection environment**

The UV ballast is suitable for protection class IP20 and its design is placed in the distribution cabinet. The cooling air must be kept clean, non-corrosive, and non-conductive. The air temperature in the power distribution cabinet must not exceed 40 °C, i.e., the cooling air must not exceed 40 °C; If the temperature in the surrounding area of the distribution cabinet exceeds the specified range, please consult the manufacturer or agent for professional advice.

2 **Recommended installation environment and conditions**

(1) Suitable for vertical or flat, do not stack one after the other.
(2) Place in the lowest temperature in the control cabinet.
(3) The distance between the ballast and the ballast is not less than 2 cm.
(4) Keep air inlets and outlets free from obstructions and obstacles.
(5) Cooling air flow for each ballast: 150m3/h.
(6) Use the supplied screws and other kits to set up and secure the controller in the power distribution cabinet.

3 **Electrical installation and connection**

According to the diagram on the ballast, use a dedicated connector to correctly connect the controller to the three-phase power supply (three-phase, 380–480 V, +6/−10%, and 48–62 Hz + PE). Three-phase faults and excessive supply voltage detection and protection have been integrated into the controller. It is recommended to use a double insurance connection.That is, in addition to grounding the power connector, a second layer of grounding insurance must be made with a $10mm^2$ copper wire at the controller's grounding bolt (M5).

When selecting the cable, it is necessary to consider the possibility of a maximum current peak value of 100 A within 50 μs of the instantaneous lighting time. The choice of cable must be considered based on the highest input current shown in the cable type table. The complete installation needs to take into account the instantaneous lighting voltage of the ballast up to 4000 V. At the same time, the lamp voltage under normal operation and the voltage under other conditions must also be considered. Ballasts have integrated the detection and protection of short circuits, open circuits, and ground fault/ground level differences. The ballast supports the series connection of several lamps.

4 **Cable size and wiring**

(1) Lighting voltage resistance 4 KV.
(2) 0 to 15A: minimum 2.5 mm^2; greater than 15A: minimum 4.0 mm^2.
(3) Lamp connection up to 50 m.
(4) The distance between the cable and the signal line must not be less than 5 cm.
(5) Use multipolar shielded cables.
(6) Do not use lamp connections to ground or any components.
(7) Do not connect the ballast to any external components (such as an external starter).
(8) Use a clamp to connect the power cable to the ballast and shield it, and then connect the controller; in addition, the cable can be bolted and fixed in the ballast using an insulated rope (Do not tie it to the lamp on the tube).

8.5 Fault Diagnosis of Typical Ballast Water Management System

Take the ballast water management systemYP≋BWMS® developed by Jiangsu University of Science and Technology as an example, please refer to the fault diagnosis Table 8.28 for the fault phenomena of the system and the possible causes and solutions of the faults.

Table 8.28 Fault diagnosis of YP≋BWMS®

Fault phenomenon	Possible reason	Solution
Ballast water management system fails to start	Switch is not closed	Close the main switch
	The power cord is not connected	Connect the power cord
Self-cleaning filter cannot self-clean	Differential pressure transmitter failure	Differential pressure transmitter replacement
	Differential pressure transmitter line blockage	Cleaning differential pressure transmitter line
	Self-cleaning control unit failure	Check the self-cleaning control unit
	Fine filter broken or blocked	Remove fine filters for new or cleaning
	Trip disc loose	Fastening trip disc coupling pin
	Proximity sensor failure	Proximity sensor replacement
	Filter backwash tube loose	Fastening filter backwash tube
	Suction branch pipe loose	Fastening suction pipe

(continued)

Table 8.28 (continued)

Fault phenomenon	Possible reason	Solution
	Sewage solenoid valve cannot be opened	Check the blowdown solenoid valve if necessary
	Sewage pipe plugging	Clear sewage pipeline
	Sewage pump failure	Check the sewage pump if necessary
	Control relay is damaged	New relay
Abnormal sound of self-cleaning filter operation	Fine mesh screen and suction pipe scraping	Adjust fine filter and suction nozzle distance to 1 ~ 2 mm
	Fine Filter and Filter Backwash Pipe Scraping	Adjust fine filter and nozzle distance to 2 mm
	Suction branch U-shaped connector loose	Fasten suction pipe U-type connector bolts
	Cleaning branch U connector loose	Fasten cleaning pipe U-type connector bolt
	Suction nozzle or nozzle loose	Fasten the suction nozzle or nozzle
Abnormal heating in the self-cleaning filter reducer	Excessive friction in drive shaft or cleaning tube	Check seals or shafts, cleaning tubes
	Reducer starts and stops too frequently	Reduce the use of reducer
	Reducer bearing wear	Repair or replace reducer bearings
	Output voltage is too high or too low	Check if the voltage is normal
Reducer vibrates too much while running	Reducer gear, bearing wear	Replacement reducer gears, bearings
	Poor fixing of reducer, loose bolts	Retighten the reducer bolt
Reducer makes too much noise while running	Big and persistent sound: bearing damage, gear wear	Replace new or professional factory repairs
	Occasionally loud: gear damage, foreign matter stuck	Replace new or professional factory repairs
High-pressure cleaning pump or sewage pump cannot start or start with large load	The prime mover or power supply is abnormal	Check power or prime mover conditions
	Pump stuck	Check the couplings manually and check the pump if necessary
	Outlet valve is not closed	Close the outlet valve and restart
	High-pressure cleaning pump or sewage pump automatic start relay failure	New relay

(continued)

Table 8.28 (continued)

Fault phenomenon	Possible reason	Solution
High-pressure cleaning pump cannot drain	Self-cleaning filter screen dirty plug	Self-cleaning filter filter
	Reverse pressure pump installation reverse	Reinstall the high pressure cleaning pump
Sewage pump cannot drain	Dirty nozzle	Clean or replace the nozzle
	Sewage pump installation reversed	Reinstall sewage pump
	Sewage pump outlet pipe plugging	Qingtong Sewage Pump Outlet Pipeline

References

1. YPBWMS®. 2013. Ballast Water Management System Operation and Maintenance Manual. Wenzhou: Zhejiang Yingpeng Marine Equipment Manufacturing Co., Ltd.
2. Li-xin, Yan. 2016. Design and research of a new self-cleaning filter. Zhenjiang: Jiangsu University of Science and Technology.
3. BSKY®. 2013. Ballast Water Management System Operation and Maintenance Manual. Wuxi: Wuxi Lantian Electronics Co., Ltd.
4. Elite®. 2016. Ballast Water Management System Operation and Maintenance Manual. Shanghai: Shanghai Hansheng Shipping Technology Co., Ltd.
5. Nie. 2014. Ballast Water Management System Operation and Maintenance Manual [R].Taizhou: Jiangsu Nanji Machinery Co., Ltd.
6. Electric diaphragm pump operating procedures[BE/OL].https://wenku.baidu.com/view/1c1dc0e6db38376baf1ffc4ffe4733687e21fcf6.html?re=view
7. Wei-hua, Chen, Li Mo-you and Wang Jin-zhou. 2005. KL-15 type oxygen generator nitrogen molecular sieve adsorber common fault and correct use. *Operation and maintenance* (12): 23–26.
8. Yu-ke, Sun. 2012. Study on High-efficiency Bactericidal System for Ship Ballast Water Treatment. Zhenjiang: Jiangsu University of Science and Technology.
9. Guozhi, Bao, Sun Yuke, and Chen Ning. 2012. Design of ship ballast water treatment device system. Zhenjiang: Jiangsu Shipping.

Chapter 9
Implementation of Ballast Water Management System on Ships

The ship ballast water management system is mainly installed in various types of ships that sail across sea. In terms of the types of ships, they are mainly bulk carriers, container ships, and oil tankers. In addition, there are marine engineering vessels that sail across the sea, scientific research cruise ships, and cruise ships.

With the imminent entry into force of the IMO Convention, there will be a large number of new ships and old ships that need to be installed or refitted with the ballast water management system. This will require the provision of implementation plans for the selection and installation of different target products for different types of ships.

For this purpose, the modification of the target ship type is required, the collection and analysis of technical data is carried out, and the technical route for installing different target products for different ship types is formulated, as shown in Fig. 9.1.

According to the two-stage processing method currently adopted for the ballast water management system target products, as shown in Fig. 9.2, and the system selection characteristics for the second processing stage currently on the market, as shown in Fig. 9.3, target product manufacturers have usually developed different flow treatment systems to meet the needs of different ship types.

9.1 The Implementation Process of Ship Modification

For modified ships, the implementation of the conversion is divided into three stages. First, use three-dimensional digital total station space scanner to collect cabin space data; secondly, the method of vectorization of lattice data is used to transfer the three-dimensional data of the cabin space collected by the modified ship to the CAD system for vectorization transformation; Third, full-scale digital stakeout is performed within its vectorized available space, so that the processing unit and piping system can be completed and tested at the company's premises, meeting the retrofitting shipowner's requirements for ship conversion cycles. Usually, the target

N. Chen et al., *Development and Implementation of Ship BWMS*,
https://doi.org/10.1007/978-981-33-6865-1_9

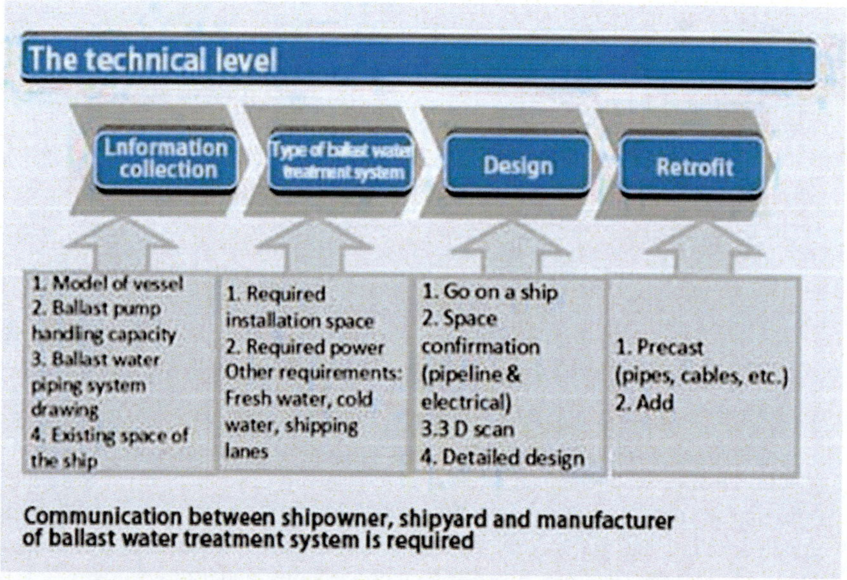

Fig. 9.1 Installation process of BWMS

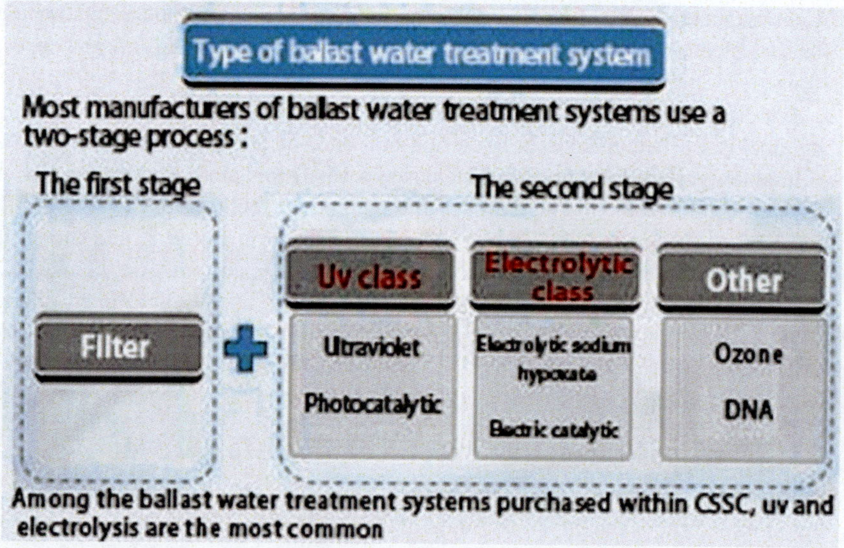

Fig. 9.2 Two-stage treatment of BWMS

classification	Ballast pump displacement (m^3/hr)	Ship type		Selection of BWMS
Ships with low dependence on BWMS	≤ 10,00	Container ship	(Most)	UV class
		Bulk carrier	Handy	
		Oil tanker	(Rare)	
Ships with general dependence on BWMS	10,00 ~ 20,00	Container ship	Ultra Large	UV & electrolysis
		Bulk carrier	Panamax	
		Oil tanker	Handy	
Ships highly dependent on BWMS	> 2000	Container ship	(Rare)	Electrolysis
		Bulk carrier	Capasize	
		Oil tanker	(Most)	

Ballast pump displacement<1,000m³/h, UV devices are dominant in space, power and price.
Ballast pump displacement>3.000m³/h. The advantage of electrolysis equipment lies in space, power and price.

Fig. 9.3 The characteristics of system selection in the second processing stage

product supplier purchases a three-dimensional digital total station space scanner, as shown in Fig. 9.4. After the target project contract takes effect, the supplier needs to collect the space data of the engine room of the refitting target ship. Figure 9.5

Fig. 9.4: 3-D digital camera space scanner

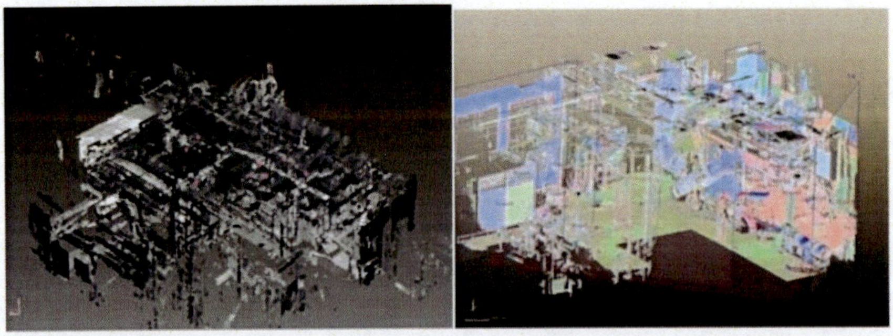

Fig. 9.5 Data of cabin space point rigid body and vectorization planning data

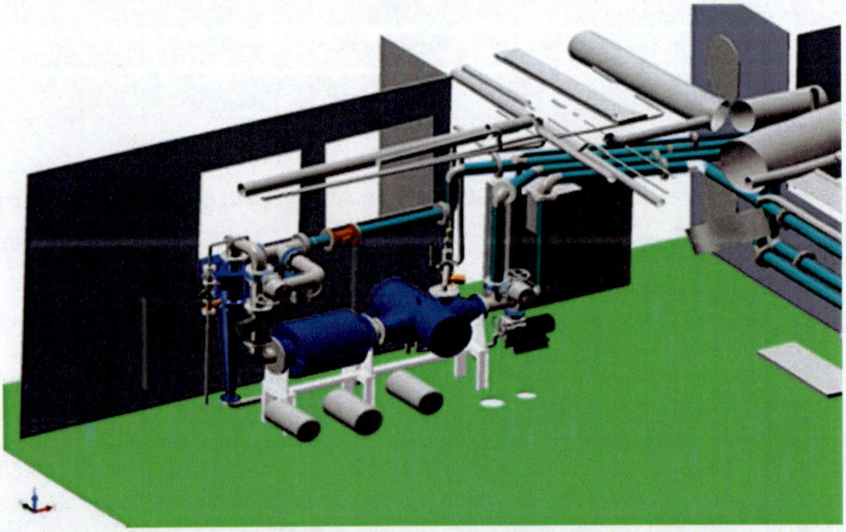

Fig. 9.6 Lofting model of system

shows the data collected by the conversion ship project and the vectorization conversion planning data. The modified partial loft design model of the project is shown in Fig. 9.6 [1].

9.2 Matters and Cases for Design of Ballast Water Management System for Modified Ship Installation

As the refitting of ships needs to adopt the technical means for rapid response to the market, the operating ships are modified in a decentralized way for different ship types and different systems. The following matters should be noted in particular while retrofitting ships to install ballast water treatment systems [2]:

(1) Whether it can be installed or only one. It mainly considers ships operating in the same sea area, such as Bohai Bay and Yellow Sea; ships without ballast water; adding this device makes the ship unable to make profit; a small installation will not affect normal operation and other factors.
(2) Consider whether the capacity of the power station is limited and whether the power grid needs to be upgraded. When selecting a processing device, it is not allowed to exceed the capacity of the power station; the size of the short-circuit current is accounted for, and the capacity of the main bus bar is determined, thereby determining whether it is necessary to adjust the main bus bar; the transformation of the power distribution screen and grid, and addition of switches, cables, etc.
(3) Consider the size of the processing system's capacity, the layout of equipment units, and the size of the pipeline's renovation project. Accurate calculations are required to consider the pressure loss caused by the treatment system to the ballast system to determine whether the ballast pump needs to be replaced; At present, many products of ballast water management systems only need to process ballast water during ballasting, and do not need to process water during the unloading phase. Due to the pressure loss of the device, the displacement of the ballast pump in the two stages of ballasting and unloading is inconsistent, so the processing capacity of the processing system is not necessarily consistent with the nominal displacement of the pump; choose a processing device with a small amount of change as much as possible; In addition, the processing device can be arranged in the engine room, empty cabin, main deck and other places; pay attention to the structure of the area where the device is located; and the optimized layout of pipelines and power cables.
(4) Modification of valve remote control system. The valves related to the ballast water treatment system are autonomously controlled by the ballast water treatment device and do not need to be connected to the valve remote control system; the ballast water treatment system and the valve remote control system need to communicate with each other, and the piping system diagram of the valve remote control system needs update.
(5) Other requirements. The pipeline material must meet the requirements of the processing equipment manufacturer; the discharge of dangerous gases must be far away from the fresh air inlet, and leak detection alarm devices must be installed at the necessary locations; during the flowmeter arrangement, one should pay attention to the requirements of the length of the straight section of the inlet and outlet pipelines; The electrical boxes, control boxes, and sensors in the area must be intrinsically safe; upstream and downstream positional relationship between sampling device and injection port, and spacing requirements.

On the basis of the above basic principles, the installation method of the operating ship system mainly considers the installation space of the ship itself, adopts a modular conversion method, appliesan overall modular approach in design, and selects the cabin unit according to the cabin space or operation characteristics. The module or deck container unit module effectively solves the problem of rapidity and short cycle time for operating ships. Especially for the old ships, due to the limitations of the cabin space, the design method of the decentralized module unit is adopted in the modification, the effective space of the engine room and the deck is fully utilized, along with the decentralized layout of the system and anoverall operation and management design method, in order to solve the old ships' conversion problems. Figure 9.7 shows the layout of a typical decentralized unit in the cabin of a bulk or container ship.

The same is true for the pharmacy method. The following example shows the modification of the pharmacy method on the 85,000 DWT ship [3] as shown in Fig. 9.8.

The modification of the pharmacy method on the 300,000 DWT VLCC oil tanker [3] is shown in Fig. 9.9.

The figure above shows that the modification of the ballast water system on the 300,000DWT VLCC oil tanker is to distribute the units of the sterilizer system in the pump room of the VLCC ship.

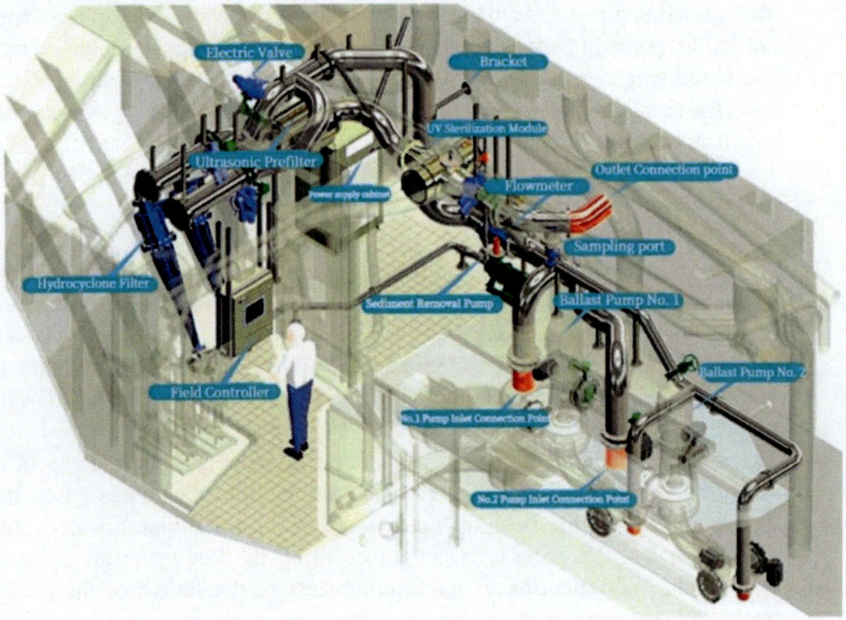

Fig. 9.7 Typical decentralized unit layout in a bulk cargo or container ship cabin

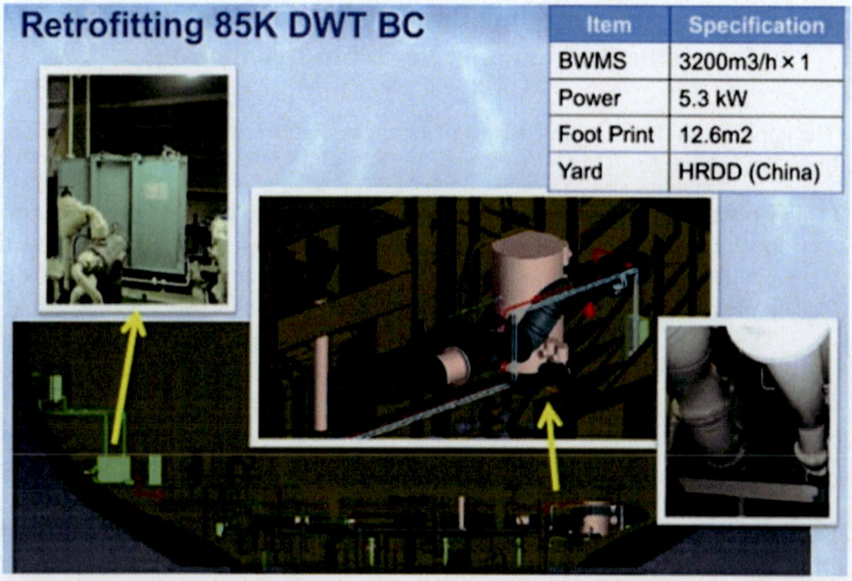

Fig. 9.8 Retrofitting 85 K DWT ship

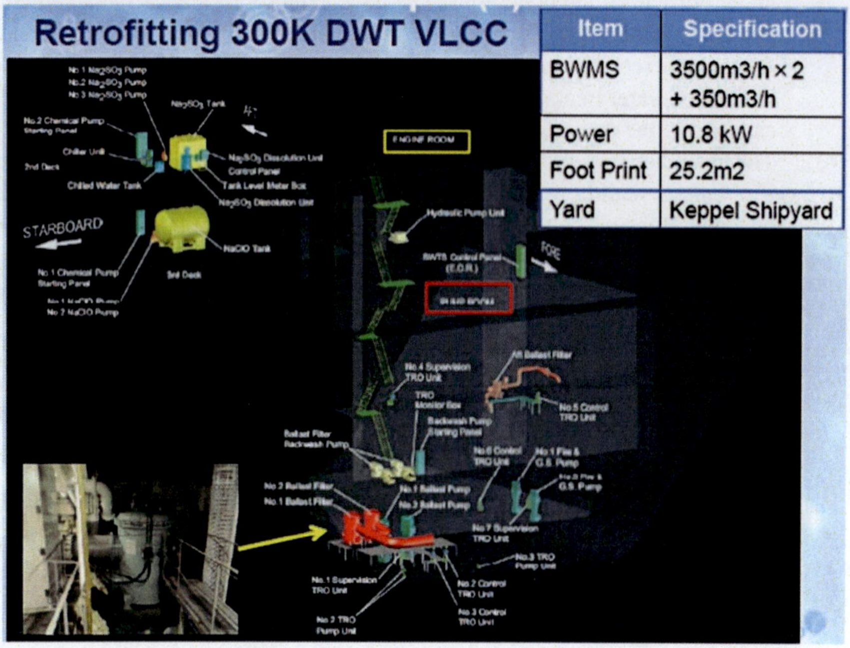

Fig. 9.9 Retrofitting 300 K DWT VLCCoil tanker

9.3 Matters and Cases for Design of Ballast Water Management System for New Ship Installation

The application of the ballast water treatment system in new ships should focus on the following aspects [2, 4]:

(1) Optimized layout of treatment equipment and ballast piping. The vast majority of the current main merchant ships are stern cabin ships. The space at the bottom of the nacelle is very cramped. At the same time, the size of the ballast piping is usually very large. If the treatment method is fully flow-through (eg, TECH CROSS) or needs filtration (dual-rule) or both (sea shield), the design of the bottom of the nacelle requires special consideration of the layout of the equipment and the direction of the pipeline to find the best solution. If there is no such feature, the traditional bottom layout of the nacelle will remain unchanged. For example, the NK-O3 and ColdHarbour's treatment system, the layout requirements for the bottom of the nacelle are relatively low.

(2) Power station restrictions. In designing, the basic electric load of the ship under normal sailing conditions and the electric load of the ballast water pump and the ballast water treatment device under the condition of loading and unloading of the dock shall be considered. The selection basis of the power station is still under normal sailing conditions. The power station is operated in an efficient mode for a long time. Underthe terminal's unloading condition, three main sets of generators simultaneously supply power to meet the requirements of the ballast water treatment device; the use of a treatment device exceeding the capacity of the power station is restricted.

(3) The characteristics of the ship type. Equipment used for oil tankers or chemical tankers shall have explosion-proof capability and the equipment may be installed in the pump room. For small, fast container liners, due to limitations in the space of the engine room, it may be considered to make the unit as a complete box-type processing unit and be installed on the side of the deck of the pilot building.

(4) Whether the ballast tank needs to maintain pressure. The ballast water treatment system in the form of inert gas needs to maintain the pressure in the ballast tanks, and its air pipe head selection needs special attention; the pressure must also be fed back to the structural professional to consider the structural design pressure.

(5) Other requirements. The material of the pipeline shall meet the requirements of the processing equipment manufacturer; the discharge of hazardous gases shall be far away from the fresh air inlet, and leak detection and alarm device must be installed at the necessary position; the arrangement of the flowmeter shall pay attention to the requirements of the length of the straight section of the inlet and outlet; within the hazardous area, the electrical boxes, control boxes, sensors etc. shall be of intrinsically safe type; upstream and downstream positional relationship between sampling device and injection port, and spacing requirements.

180000DWT bulk carrier real ship application case, the system principle is shown in Fig. 9.10 and the system layout model is shown in Fig. 9.11.

113,000 COT crude oil wheel real ship application case, the system principle is shown in Fig. 9.12 and the system layout model is shown in Fig. 9.13.

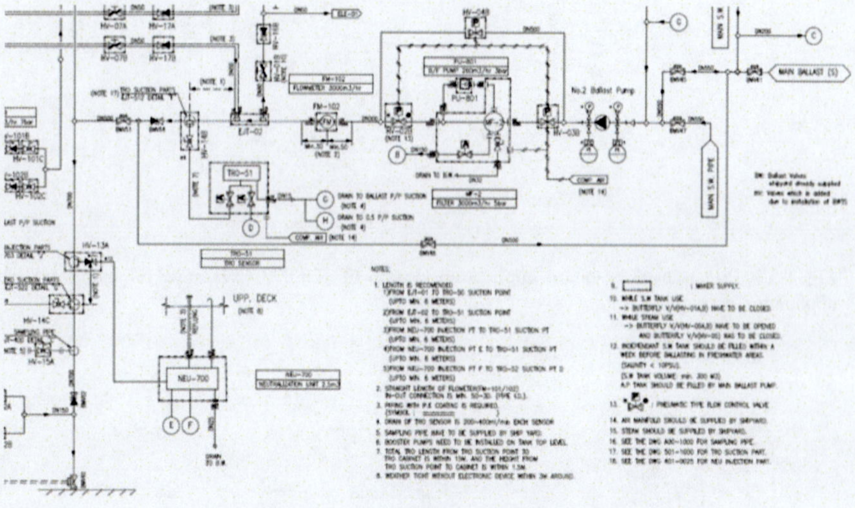

Fig. 9.10 Principle of on-board application system of a 180,000DWTbulk carrier

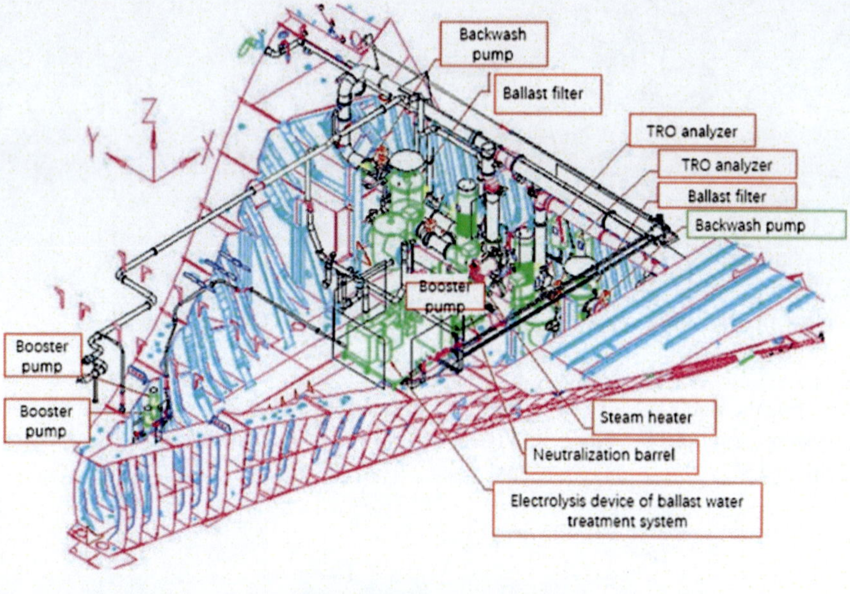

Fig. 9.11 The lofting model of a 180,000DWT bulk carrier

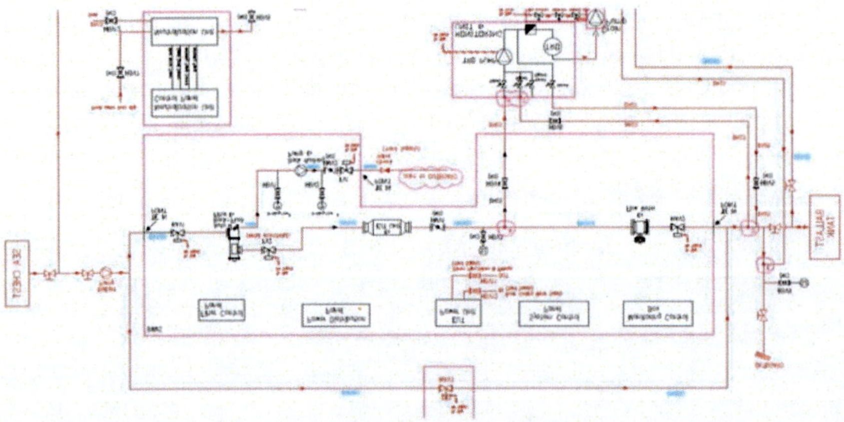

Fig. 9.12 Principle of on-board application system of 113,000 COTcrude oil wheel real ship application case

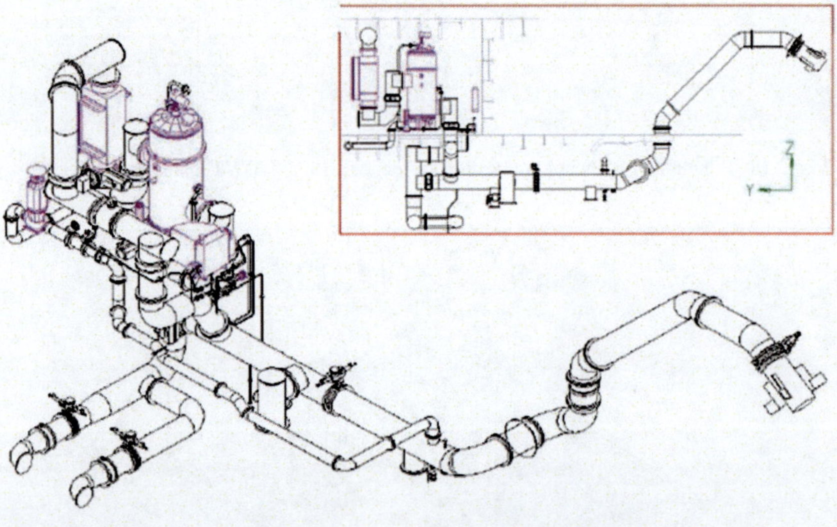

Fig. 9.13 The lofting model of 113,000 COTcrude oil wheel real ship application case

For the 85000VLGC large liquefied gas carrier actual ship application case, the system principle is shown in Fig. 9.14 and the system setup model is shown in Fig. 9.15.

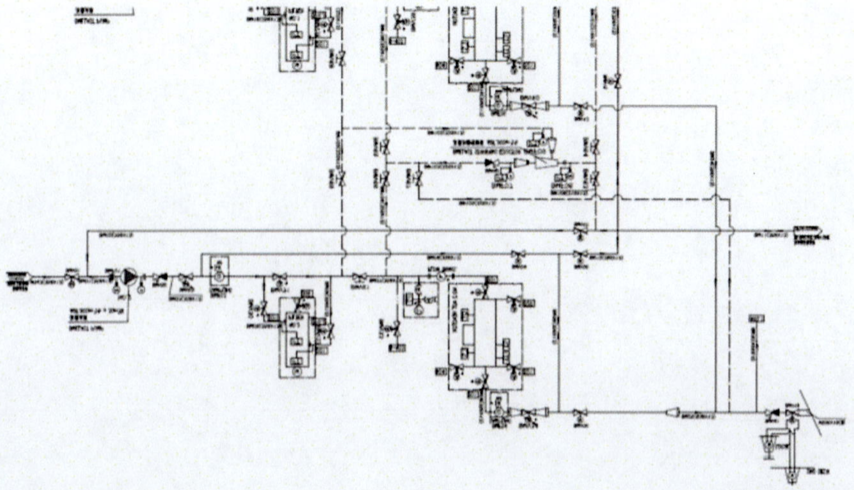

Fig. 9.14 Principle of on-board application system of 85,000 VLGClarge-scale liquefied gas carrier

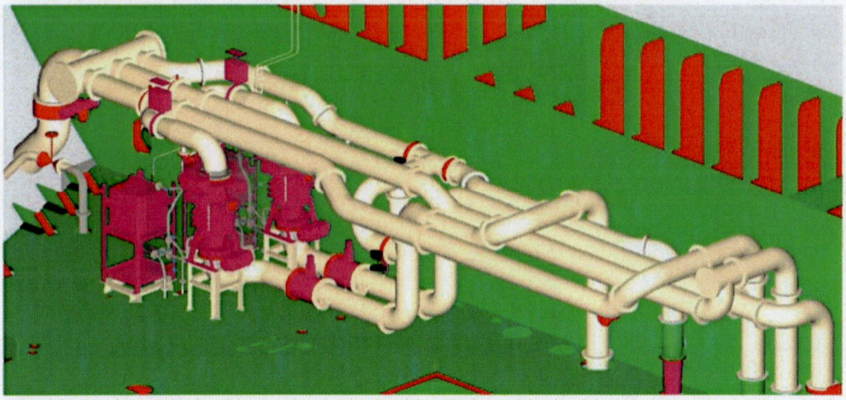

Fig. 9.15 The lofting model of 85,000 VLGClarge-scale liquefied gas carrier

References

1. BSKY. 2013. Ballast Water Management System Operation and Maintenance Manual. Wuxi: Wuxi Lantian Electronics Co., Ltd.
2. Pengfei, Wu. 2016. Application solution of ballast water treatment system. Shanghai: 2016 Annual Meeting of High Power Marine Diesel Engine and Ballast Water Management System.
3. Electric diaphragm pump operating procedures [BE/OL].https://wenku.baidu.com/view/1c1dc0e6db38376baf1ffc4ffe4733687e21fcf6.html?re=view
4. Linxing, Wang, Wu Zonghuan and Chen Ning. 2017. R&D and Industrialization Technology Report of High-efficiency Combined Physical Treatment Ship Ballast Water Treatment Device. Wuxi: Wuxi Lantian Electronics Co., Ltd.

Postscript

This book was completed in a self-help manner with the support of the following funding projects: (1) Jiangsu Province Science and Technology Achievements Transformation Special Fund Project (BA2013065) "R&D and industrialization of high-performance composite physical treatment ship ballast water treatment devices"; (2) a forward-looking joint research project of the Ministry of Science and Technology of China (BY2015065-10) "Research on energy-saving technologies for ship ballast water treatment systems of UVC + US"; and (3) the emerging industry project of Zhenjiang City (CY2010019) "Development and research on marine ballast water marine biological treatment devices".

Based on a large number of independent design, experimental and simulation optimization data, the book collects the data of the more mature ballast water management system products at home and abroad, and summarizes the best practices for the design and optimization of system unit modules, related to the development and implementation of ship ballast water management system, along with the precautions regarding system operation and maintenance, old ship modifications, and new ship design.

In the process of writing this book, several graduate students, including YukeSun, Chen Jiao, Xianzheng Meng, Meng Meng, and Zhiming Wang, did lots of design and testing work. Professor Guozhi Bao participated in the development of the control system. The data of other relevant products mentioned in the book was obtained with the help and support of LinxingWang, general manager of Wuxi Lantian Electronics Co., Ltd., Qiming Wang, engineer of Shanghai Hansheng Shipbuilding Technology Co., Ltd., Jiansheng Bi, deputy chief engineer of Bohai Shipbuilding Industry Group, Pengfei Wu, engineer of Shanghai Pengqiao Shipbuilding Co., Ltd.,, Liqun Tian, deputy director of the intelligence unitin Shanghai 704th Research Institute, and Lei Wang, director of Shanghai Shipbuilding Technology Research Institute. We thank them all for their help and support.

N. Chen et al., *Development and Implementation of Ship BWMS*,
https://doi.org/10.1007/978-981-33-6865-1

The ballast water management system covered in this book has a wide range of technical aspects, and it is inevitable that there are deficiencies. We hope to receive comments and advice from the readers so that future improvement to the book can be made.

Authors

June 23, 2019.

Appendix 1
Sound Pressure Simulation Program

```
L=0.640;
W=0.280;
H;
a=0.067/2;
f;
d1=0.080;
d2=0.070;
N=8;
M=4;
c=1450;
Lamda=c/f;
w=2*pi*f;
K=2*pi/Lamda;
rho=1040;
K´=rho*f*u0;
n=20;
k=0;
j1=0;
j2=0;
j3=0;
```


N. Chen et al., *Development and Implementation of Ship BWMS*,
https://doi.org/10.1007/978-981-33-6865-1

```
for x=-1*L/2:L/(n-1):L/2
    j1=j1+1;
    j2=0;
    j3=0;
for y=-1*W/2:W/(n-1):W/2
      j2=j2+1;
      j3=0;
for z=0:H/(n-1):H
         j3=j3+1;
         Alfa=atan(y/x);
         Theta=atan(x/(z* sin(Alfa)));
         r0=z/cos(Theta);
Rax1=abs((sin(2*pi*a*N*d1/Lamda*(sin(Theta))*cos(Alfa))+eps)./(N*sin(2*pi*a
*d1/Lamda*(sin(Theta))*cos(Alfa))+eps));
Rax2=abs((2*besselj(1,2*pi*a/Lamda*sin(Theta))+eps)./(2*pi*a/Lamda*sin(Thet
a)+eps));Rax3=abs((sin(2*pi*a*M*d2/Lamda*(sin(Theta))*cos(Alfa))+eps)./(M*s
in(2*pi*a*d2/Lamda*(sin(Theta))*cos(Alfa))+eps));
Rax4=abs((2*besselj(1,2*pi*a/Lamda*sin(Theta))+eps)./(2*pi*a/Lamda*sin(Thet
a)+eps));
Rax=Rax1.*Rax2.*Rax3.*Rax4;
%Lower transducer array directivity function
P0x=M*N*K´*pi*a^2/r0;
Theta0=atan(r0*sin(Theta)/(H-r0*cos(Theta)));
Ras1=abs((sin(2*pi*a*N*d1/Lamda*(sin(Theta0))*cos(Alfa))+eps)./(N*sin(2*pi*
a*d1/Lamda*(sin(Theta0))*cos(Alfa))+eps));
Ras2=abs((2*besselj(1,2*pi*a/Lamda*sin(Theta0))+eps)./(2*pi*a/Lamda*sin(The
ta0)+eps));
Ras3=abs((sin(2*pi*a*M*d2/Lamda*(sin(Theta0))*cos(Alfa))+eps)./(M*sin(2*pi*
a*d2/Lamda*(sin(Theta0))*cos(Alfa))+eps));
Ras4=abs((2*besselj(1,2*pi*a/Lamda*sin(Theta0))+eps)./(2*pi*a/Lamda*sin(The
ta0)+eps));
Ras=Ras1.*Ras2.*Ras3.*Ras4;
```

```
%Upper transducer array directivity function
P0s=M*N*K´*pi*a^2/((H^2+r0^2-2*H*r0*cos(Theta))^0.5);
P(j1,j2,j3)=P0x.*Rax+P0s.*Ras;
%Sound pressure
I(j1,j2,j3)=(P(j1,j2,j3))^2/(2*(rho)*c);
%Sound intensity


end
end
end
[X,Y,Z]=meshgrid(linspace(-L/2,L/2,n),linspace(-
W/2,W/2,n),linspace(0,H,n));
xslice=2*L/3;
yslice=2*W/3;
zslice=[H/5,2*H/5,3*H/5,4*H/5,H];
figure(1);
slice(X,Y,Z,P,xslice,yslice,zslice);
figure(2);
slice(X,Y,Z,I,xslice,yslice,zslice);
[X,Y]=meshgrid(linspace(-L/2,L/2,n),linspace(-W/2,W/2,n));
z=H/2;
figure(3);
surf(X,Y,P);
figure(4);
surf(X,Y,I);
P(P==inf)=0;
PP=unique(P(~isnan(P)));
Pam=mean(PP(:));
NU1=sum(abs(abs(PP)-Pam))/numel(PP)/Pam;
```

Appendix 2
Simulation Program for UV Light Intensity and Dose

1. Coordinate output program

```
#define a 16

#define b 18

#define PATH  "d:\\d.txt"

#define AA 20000

int main(void)

{

       double x0, y0, x, y;

       double total_length[AA];

       FILE *fp;

       fp = fopen(PATH, "w+");

       x0 = -1 * a;

       y0 = -1 * b;

       int i = 0;

       if (!fp)

       {

              printf("open file error !\n");

       }
```


N. Chen et al., *Development and Implementation of Ship BWMS*,
https://doi.org/10.1007/978-981-33-6865-1

```
        for(x=x0; x <= a; x += 0.45)
        {
                for(y = y0; y <= b; y += 0.45)
                {
                double r1 = sqrt((x-14)*(x-14) + (y-4)*(y-4));
                double r2 = sqrt((x-7)*(x-7) + (y-12)*(y-12));
                double r3 = sqrt(x*x + (y-4)*(y-4));
                double r4 = sqrt((x+14)*(x+14) + (y-4)*(y-4));
                double r5 = sqrt((x+7)*(x+7) + (y-12)*(y-12));
                double r6 = sqrt((x+14)*(x+14) + (y+4)*(y+4));
                double r7 = sqrt(x*x + (y+4)*(y+4));
                double r8 = sqrt((x-14)*(x-14) + (y+4)*(y+4));
                double r9 = sqrt((x+7)*(x+7) + (y+12)*(y+12));
                double r10 = sqrt((x-7)*(x-7) + (y+12)*(y+12));
                total_length[i++] = r1 + r2 + r3 + r4 + r5 + r6 + r7 + r8 + r9
+ r10;
        If((r1>1.6)&&(r2>1.6)&&(r3>1.6)&&(r4>1.6)&&(r5>1.6)&&(r6>1.6)&&(r7>1
.6)&&(r8>1.6)&&(r9>1.6)&&(r10>1.6))
        fprintf(fp,"%f %f\n",x,y);
                fprintf(fp,"%f %f %f %f %f %f %f %f %f
%f\n",r1,r2,r3,r4,r5,r6,r7,r8,r9,r10);
                }
        }
        printf("%d\n",i);
        fclose(fp);
        return 0;
```

2. Light intensity and dose calculation program

```
pi=3.1415926;
P=400000;
L=25.8;
```

```
a=0.2326;
R0=1.6;
t=0.613;
I1=0.99*0.9*0.5*P*exp(-a*(r1-R0))./(pi*r1*L);
I2=0.99*0.9*0.5*P*exp(-a*(r2-R0))./(pi*r2*L);
I3=0.99*0.9*0.5*P*exp(-a*(r3-R0))./(pi*r3*L);
I4=0.99*0.9*0.5*P*exp(-a*(r4-R0))./(pi*r4*L);
I5=0.99*0.9*0.5*P*exp(-a*(r5-R0))./(pi*r5*L);
I6=0.99*0.9*0.5*P*exp(-a*(r6-R0))./(pi*r6*L);
I7=0.99*0.9*0.5*P*exp(-a*(r7-R0))./(pi*r7*L);
I8=0.99*0.9*0.5*P*exp(-a*(r8-R0))./(pi*r8*L);
I9=0.99*0.9*0.5*P*exp(-a*(r9-R0))./(pi*r9*L);
I10=0.99*0.9*0.5*P*exp(-a*(r10-R0))./(pi*r10*L);
I=I1+I2+I3+I4+I5+I6+I7+I8+I9+I10;
%Light intensity
D=I*t;
%dose

fid=fopen('I.txt','wt');
fprintf(fid,'%g\n',I);
%Output light intensity file
fclose(fid);
fid=fopen('d.txt','wt');
fprintf(fid,'%g\n',D);
%Output dose file
scatter(X,Y,5,I);
scatter(X,Y,5,D);
%Draw light intensity and dose plane map
```

```
mx=min(X); Mx=max(X);

my=min(Y); My=max(Y);

Nx=5435; Ny=5435;

cx=linspace(mx,Mx,Nx);

cy=linspace(my,My,Ny);

cI=griddata(X,Y,I,cx,cy','cubic');

cD=griddata(X,Y,D,cx,cy','cubic') ;

mesh(cx,cy,cI,'LineWidth',2) ;

mesh(cx,cy,cD,'LineWidth',2)

%Draw a three-dimensional map of light intensity and dose

[C,h]=contour(cx,cy,cD,5);

[C,h]=contour(cx,cy,cD,5)

%Draw light intensity and dose contour map

clabel(C,h) xlabel( 'x /cm' );

ylabel( 'y /cm' );

zlabel( 'I /mw/cm^2'
```

References

1. Hallegraeff, G.M., and D.J. Bolch. 1992. Transport of diatom and dinoflageuate resting spores in ships bauast water: Implications for plankton biogeography and aquaculture. *Plankton Res.* 14 (8): 1067–1084.
2. Jinjuan, Zhou. 2012. Structural Design and Simulation of High Output Ultrasonic Transducer. Nanjing:Master thesis, Nanjing University of Aeronautics and Astronautics, 14–15.

N. Chen et al., *Development and Implementation of Ship BWMS*,
https://doi.org/10.1007/978-981-33-6865-1